AF765619

The Library | UNIVERSITY OF BATH

Bath BA2 7AY · United Kingdom

NOT FOR LOAN

1? DEC 1997

16 JAN 1998

21 APR 2004

12 JAN 2000

20 JAN 2000

13 APR 2005

29 SEP 2000

23 NOV 2000

UNIVERSITY OF BATH LIBRARY

- 1 FEB 2022

Withdrawn

Mind, Brain, and the Environment

Mind, Brain, and the Environment

The Linacre Lectures 1995–6

Edited by

BRYAN CARTLEDGE
Former Principal of Linacre College
University of Oxford

Oxford New York Tokyo
OXFORD UNIVERSITY PRESS
1998

Oxford University Press, Great Clarendon Street, Oxford OX2 6DP
Oxford New York
Athens Auckland Bangkok Bogota
Bombay Buenos Aires Calcutta Cape Town
Dar es Salaam Delhi Florence Hong Kong
Istanbul Karachi Kuala Lumpur
Madras Madrid Melbourne Mexico City
Nairobi Paris Singapore Taipei
Tokyo Toronto Warsaw
and associated companies in
Berlin Ibadan

Oxford is a trade mark of Oxford University Press

Published in the United States
by Oxford University Press Inc., New York

© Oxford University Press, 1998

All rights reserved. No part of this publication may be reproduced, stored in a retrieval system, or transmitted, in any form or by any means, without the prior permission in writing of Oxford University Press. Within the UK, exceptions are allowed in respect of any fair dealing for the purpose of research or private study, or criticism or review, as permitted under the Copyright, Designs and Patents Act, 1988, or in the case of reprographic reproduction in accordance with the terms of licences issued by the Copyright Licensing Agency. Enquiries concerning reproduction outside those terms and in other countries should be sent to the Rights Department, Oxford University Press, at the address above.

This book is sold subject to the condition that it shall not, by way of trade or otherwise, be lent, re-sold, hired out, or otherwise circulated without the publisher's prior consent in any form of binding or cover other than that in which it is published and without a similar condition including this condition being imposed on the subsequent purchaser.

A catalogue record for this book is available from the British Library

Library of Congress Cataloging in Publication Data
(Data available)

ISBN 0 19 854992 X

Typeset by Hewer Text Compostion Services, Edinburgh
Printed in Great Britain by
Biddles Ltd, Guildford & King's Lynn

UNIVERSITY OF BATH
LIBRARY
3 18 NOV 1997
159.937.4
5110484

Acknowledgements

I am glad to record, for the third time, the gratitude of Linacre College, and of Oxford University, to British Petroleum plc for their generosity in sponsoring the Linacre Lectures. Taken as a whole the Linacre Lectures, of which this is the sixth series, have a global perspective; it is, therefore, appropriate that the College should have carried this project forward in partnership with BP, a company whose interests are also global and whose activities demand an enlightened environmental approach. We have enjoyed working together.

I am grateful, once more, to Frances Morphy for giving her time and expertise to the task of converting these Lectures into a book; her help throughout the six years of the project has been indispensable. Linacre's College Secretary, Jane Edwards, has once again played an invaluable role in helping me to organize the lectures themselves and in acting as a totally reliable trouble shooter.

My final vote of thanks goes to my wife, Freda Newcombe, who suggested the theme for this series of lectures and helped me to assemble an exceptionally distinguished cast of contributors.

Linacre College, Oxford
August 1996 B.G.C.

Contents

Authors

Professor Colin Blakemore, PhD, ScD, FRS
Waynflete Professor of Physiology, University of Oxford

Sir Bryan Cartledge, KCMG, MA
Former Principal, Linacre College, Oxford, 1988–96

Professor Anthony W. Clare, MD, FRCPI, FRCPsych, MPhil
Medical Director, St Patrick's Hospital, Dublin

Professor Antonio R. Damasio, MD, PhD
Van Allen Professor and Head of Department of Neurology, University of Iowa

Professor Hugh Freeman, FRCPsych, MA, MSc, FFPHM
Honorary Consultant Psychiatrist to the Salford Health Authority and Honorary Visiting Fellow, Green College, Oxford

Professor Robert A. Hinde, CBE, FRS
Former Royal Society Research Professor, University of Cambridge

Professor Tim Ingold, PhD
Max Gluckman Professor of Social Anthropology, University of Manchester

Professor David Marsden, DSc, FRCP, FRS
Professor of Clinical Neurology, National Hospitals for Neurology and Neurosurgery, London

Professor Andy Young, PhD, DSc
Senior Scientist, MRC Applied Psychology Unit, Cambridge

Introduction

Bryan Cartledge

On 3 May 1995, the eve of the fiftieth anniversary of VE Day, some two hundred people gathered in the magnificent gardens of St Hugh's College, Oxford, to enjoy tea, sunshine, and the music of the band of the Light Infantry. Among them were 35 elderly men who, although coming from widely differing social and occupational backgrounds, had a great deal in common. They were all veterans of World War II; they had all sustained serious head injuries while on active service; they had all been treated in the St Hugh's Military Hospital whose hutments had covered the gardens during and until shortly after the war; and they had all benefited from continuous contact, over half a century, with the Head Injury Advice Bureau which Professor Ritchie Russell, the eminent neurologist, had established in Oxford before the war ended. The occasion was a reunion of comrades in arms who had successfully overcome a common misfortune; and also marked the formal deposit in the St Hugh's College Library of the invaluable medical records which Ritchie Russell's Bureau had assembled over the years. It was the latest episode in a remarkable Oxford story which has direct relevance to the themes of these Linacre Lectures on 'Mind, Brain, and the Environment'.

In World War I, head injuries caused by penetration of the brain by bullets or shrapnel were usually fatal: it has been calculated that up to 1917 90% of the victims of such wounds died. In World War II, over 90% of British servicemen with head injuries not only survived but in many cases recovered sufficiently to go on to lead normal and productive lives. This extraordinary turn round was due very largely to the work of Oxford's first Professor of Surgery, Hugh (later Sir Hugh) Cairns. Professor Cairns not only persuaded the War Office to establish a significant number of mobile neurosurgical units which could work in close proximity to the front line, carry out emergency treatment, and arrange for the rapid evacuation of the head-injured; he also trained the personnel of these units and set up the Military Hospital in St Hugh's College to which the casualties were evacuated and in which they benefited from early and expert treatment. Over one thousand head-injured servicemen passed through St Hugh's before May 1945.

Oxford's involvement did not cease with their treatment and discharge. Professor Russell was quick to appreciate the immense scientific value of past and, more importantly, future case histories of the St Hugh's patients; and the importance to them of a continuing source of advice and help. The team which he built up around the Head Injury Advice Bureau thus had a dual function: rehabilitation and research. The study, on the basis of both questionnaire and periodic interview, of the capacities, behaviour, and occupational development of the St Hugh's veterans yielded valuable information about the organization of the human brain. The relationship between damage to a particular part of the brain and a specific physical or mental limitation was the subject of a succession of scientific papers which threw new light on the brain's modular structure. In 1963, Professor Russell brought a psychologist, Freda Newcombe, to Oxford to conduct a neuropsychological follow-up study of a representative sample of former St Hugh's patients. One hundred and fifty-seven veterans were interviewed and asked to carry out a number of tests: they did so willingly, both then and in subsequent years, regarding their contribution as a small repayment for their treatment in St Hugh's and subsequent rehabilitation.

The follow-up study, aspects of which were continued and deepened by Dr Newcombe and her research colleagues during the following thirty years, revealed, in addition to a wealth of specialized neuropsychological data, a remarkable picture of human resilience and sheer grit. Among those former patients who had sustained injury to only one side of the brain, over 80% had resumed employment at or above their pre-injury level; even among those with more serious, bilateral injuries the proportion was two-thirds. The former patients included a surgeon, a distinguished judge, and a national finalist in The Times crossword puzzle competition. Despite selective and specific limitations, whether of speech, vision, or movement, the great majority have led relatively normal and certainly rewarding lives, untroubled by psychiatric problems of any kind, and have matched or exceeded normal life expectancy.

Similar studies in the United States of head-injured American veterans of the Vietnam War reveal a strikingly different picture. Among them, the percentage of those who have resumed employment at pre-injury level, or any employment at all, is low. The percentage of those with psychiatric or behavioural problems is high. There have been a significant number of suicides and a considerably higher than normal level of antisocial activity leading to prosecution. To a lesser extent – but to a significantly greater extent than American World War II veterans – the same is true of US servicemen who sustained head injuries while fighting in Korea. Why should this be so?

The probable answer brings us close to some of the themes which are developed in this series of Linacre Lectures. Several lecturers explore the two-way relationship between the brain or mind and its environment. In Robert Hinde's words, 'we shall not fully understand the impact of the environment on our behaviour, or of our behaviour on the environment, unless we come to terms with the dialectical relations between them'. Hugh Freeman writes of 'the urgent need to examine those aspects of people's physical and social surroundings which are likely to influence mental health'. War is an extreme example of the dialectical relationship which Hinde describes. A decision to wage war, whether it springs from fear or from greed, and the prosecution of that decision at every level are destructive of both the physical and the social environment; the destruction is inevitably compounded by the defensive efforts and struggle to survive of those who have war thrust upon them. The environment which the combatants have created is itself malign, inflicting death, injury, and intolerable levels of stress.

The capacity of an individual to come to terms with injury, whether to the body or the brain, and high levels of stress – whether induced by fear, anger, or grief – depends in part upon that individual's perception of himself in relation both to his physical and to his social environment. Allied servicemen in World War II had a clear perception of the rationale for their exposure to danger and hardship: they were fighting to defend their families and their homelands against invasion and occupation. They believed, more or less passionately, in that cause and their societies wholeheartedly endorsed it, according them honour, praise, and gratitude. The perception of US servicemen fighting in Vietnam of the rationale for their presence there was much more confused. The 'domino theory', which predicated the loss of the whole of South-East Asia to Communism if the Viet Cong prevailed in Vietnam, was a hypothesis which lacked immediacy and carried little conviction while the war was in progress, however much it was justified by events after the US withdrawal. To many American servicemen, the war was being fought not to defend the United States nor even American interests, but a venal and corrupt regime in a distant country where Americans had no business to be. This confusion about the rights and wrongs of the war was shared by American society as a whole. Returning veterans of the Vietnam war, whether wounded or whole, could by no means be sure of a hero's welcome: American civilians tended to turn their backs on the sacrifices and suffering of the war's victims, who represented unwelcome reminders of an episode in their country's history which was increasingly regarded as a shameful humiliation, best forgotten. (Twenty years later, Soviet veterans of the war in Afghanistan, the *afgantsy*, had to endure a similar experience; although medical data are not publicly available, it seems likely that rehabilitation of the head-injured suffered in consequence.)

It is probable that the contrast between the grateful, supportive social environment in which the recovery and rehabilitation of the British head-injured veterans took place and the less sympathetic social climate with which American head-injured veterans of Vietnam had to contend had a great deal to do with the rather remarkable contrast between their respective group case histories. The types of injury which the two groups sustained were similar or identical; and the surgical or medical treatment which they received can be assumed to have been at least on a par – indeed, one would expect the veterans of Vietnam to have benefited from advances in surgical techniques and antibiotics during the two decades between the end of World War II and the war in Vietnam. Nevertheless, both the physical and the psychological rehabilitation of the World War II group was significantly more successful than that of the Vietnam group.

The environments in which the two groups of servicemen both sustained and embarked on recovery from their head injuries differed in another important respect. Hallucinogenic drugs were virtually absent from the social culture of the 1940s and 1950s. By the late 1960s, they were commonplace. Their use was widespread among US servicemen in Vietnam and was becoming so in the society to which they returned for treatment and rehabilitation; addictions acquired to insulate the senses from the stress and horror of a peculiarly nasty war were all too easy to sustain after evacuation from the theatre of operations to US Pacific bases or the US mainland. In his contribution to this volume (pages 107–123), David Marsden refers to the hazards to which so-called 'recreational' drugs expose the human brain. Moreover, he gives a graphic example of the disastrous neurological damage which can be inflicted by 'designer' drugs with a design fault: such drugs are all too likely to have added to the risks to which those head-injured Vietnam veterans who became addicted were exposed, and to which the nature of their injuries rendered them particularly vulnerable. Casualties of the environment of war, many of them also became casualties of an unsupportive civilian environment in which it was easier to succumb to their disabilities than to overcome them.

This comparative sketch of two groups of men who shared similar afflictions but in significantly different social environments is just one example of the interaction between mind, brain, and environment; the chapters which follow, based on the 1995/96 series of Linacre Lectures, contain many more. The contributors address the subject from a wide range of scientific perspectives: together, they demonstrate the remarkable progress made by modern neuroscience in unravelling a complex dialectical relationship.

ACKNOWLEDGEMENT

This introduction draws substantially on an article by Dr Freda Newcombe, 'Very late outcome after focal wartime brain wounds', in the *Journal of Clinical and Experimental Neuropsychology* (1996), **18** (1), 1–23; the same arguments were concisely summarized in her letter to The Times of 10 June 1988 on the aftermath of wartime head injuries, in the context of the Falklands War.

1
Humans and human habitats: reciprocal influences

Robert A. Hinde

Professor Robert Hinde, CBE, FRS, spent four years of the Second World War serving as a Coastal Command pilot in the RAF and then took a first degree at Cambridge. He obtained his DPhil in 1950 at Oxford's Edward Grey Institute of Field Ornithology. Robert Hinde then returned to Cambridge as Curator of the Ornithological Field Station (later sub-Department of Animal Behaviour) at Madingley and soon after was elected to a Research Fellowship of his old College, St John's. The College and Madingley have been his academic homes ever since. He served at different times as Steward, Tutor, and Master (1989–1994) in his College and conducted research on animal and human behaviour at Madingley. For the last twenty years his main interest has been in human personal relationships. He is an Honorary Foreign Associate of the (US) National Academy of Sciences, an Honorary Fellow of Balliol College, Oxford and of the Royal College of Psychiatry. His contributions have been recognized by academic bodies concerned with anthropology, developmental psychology, ethology, ornithology, personal relationships, primatology, and psychiatry; he holds a number of honorary doctorates. In addition to his scientific interests, Robert Hinde has written with authority on issues of peace and war and on nuclear disarmament.

We are all rightly concerned about the effects of human activities on the environment. The pollution of the atmosphere, the destruction of the ozone layer, the poisoning of the seas and rivers, the steady erosion of biodiversity, and the depletion of unrenewable resources; these are all consequences of what we are doing to the earth. In due course they will rebound on our heads and on those of our descendants unless we change course. Human activities affect the environment in diverse ways; some are beneficial and others may be beneficial to some but are deleterious to many. Thus it is important to attempt to tease apart the relations between human behaviour,

our experience in the environment, the effects of our behaviour on the environment, and the further effects of the environment on us. I shall return briefly to these issues later, but my main theme is that if we are to understand what we are doing, we must come to terms with the fact that we affect our environment in ways that are a direct consequence of what we are.

I have just written of the effects of our behaviour on the environment and of the environment on us – and I wonder whether you, the reader, objected to the words that I used. For such words imply a dichotomy between the environment and us. They refer to an environment 'out there', separate from us, and that is exactly the view that I try to counter in this chapter. To a degree, and in ways that I shall discuss, we are continually creating our environment and our environment is continually creating us. In a sense I am anticipating a theme from Chapter 2 by Colin Blakemore: writing of brain development he describes how the cerebral cortex induces the growth of fibres from the thalamus, and how those fibres from the thalamus guide the differentiation of the cortex. In such a situation we can see the cortex as environment to the thalamus, or the thalamus as environment to the cortex. I am pursuing this theme by focusing attention on the dialogue, or dialectical relations, between humans and their environment. I shall suggest that an approach which presupposes a hard distinction between individual and environment obscures two important issues. The first concerns the distinction between the environment that we can experience and whatever else is 'out there' which we cannot experience. What we cannot experience is not properly considered part of our environment. (I neglect the possibility that we may be affected by influences that we cannot detect, like strong electric fields.) I shall consider this distinction first and briefly. Second, and perhaps of more interest, is the environment that we do experience: here I argue that what we experience is a product of what we are and do, and that we need to come to terms with a continuing dialectic between ourselves and the environment that we experience.

THE ENVIRONMENT THAT WE CAN EXPERIENCE

To some aspects of the world 'out there' our sense organs are responsive, but other aspects just pass us by. The house that we experience is very different from that experienced by the cat who shares the same physical structure with us. We are relatively insensitive to the myriad odours that she experiences, and we see a world of coloured form more variegated than she can experience. The flowers we see in the garden are very different from those seen by the bee which visits them: we cannot see the ultraviolet that she can. The sounds we hear in the twilight are very different from those the bat hears and makes, most of which we cannot even detect. Our sense organs do not respond to some of the stimuli to which cat, bee, and bat respond, and ours respond to some that theirs do not: since we have all grown up in the same physical world we could say that our genes determine the differences between our environments.

But while within an individual lifetime it can be said that the genes determine one's environment, on an evolutionary time-scale the physical world has determined, or at any rate influenced, our genes. The sense organs of different species have become specialized to respond to particular aspects of the physical world in accordance with their ways of life. For example, the sound frequencies used by whales, humans, and bats occupy bands of increasing frequency, each suited to the ecology of the species. The very low frequencies used by whales carry for very long distances in the oceans, the high frequencies used by bats are suitable for the echo-location of obstacles and prey.

Furthermore, the mechanisms by which we perceive the stimuli impinging on our sense organs have subtleties of which we are not immediately aware. Consider how we perceive colour (Shepard 1992). In the daytime, the light that enters our eyes comes from the sun. However, the nature of the light coming from a particular object depends critically on the conditions. If the sun is directly overhead, it is likely to be relatively strong in middle wavelengths (yellows), but if the light comes from a clear sky, or from the setting sun, it may be richer in short (blue) or long (red) wavelengths respectively. Yet we perceive the object as having a relatively constant colour. It is important that we should, because to respond to it appropriately we must see it as the same object even though the light reaching our eyes changes markedly in spectral composition. This remarkable feat of colour constancy depends on an ability to separate the contributions to the stimuli we perceive – of general lighting on the one hand and of the surface of the object on the other. This is apparently achieved by a mechanism which transforms the spectral characteristics of the light entering the eye in accordance with the spectral characteristics of the reflecting surface (Maloney and Wandell 1986).

Turning to another aspect of colour perception, although solar light involves a continuum of wavelengths, we perceive it categorically, as divided into a few relatively discrete bands that we label with such terms as red, orange, yellow, green, blue, and violet. The prototypical colours are more rapidly perceived and remembered than the intermediates, and discrimination of colours tends to be sharper across the boundaries of these colour regions than within them. This is not just a matter of our having common language terms for the so-called colours of the rainbow. Perceptual categorization of colours is found also in other species, though not necessarily with the same boundaries. Just what its biological function may be is not clear, but amongst human beings living in different parts of the world, with different languages and colour terms, the categories seem to be at least very similar. And although the many languages studied by Berlin and Kay (1969) differed in the number of basic colour terms that they included, they conformed to an order of occurrence of colour categories:

white and black > red > green and yellow > blue > brown > purple, pink, orange and grey,

where $a > b$ indicates that *a* is present in every language in which *b* is present and also in some language(s) in which *b* is not.

The importance of red is particularly interesting. For nearly a third of the languages, red was the only colour word in addition to black and white. Humphrey (1983) has demonstrated experimentally that red is aversive to rhesus monkeys, and has reviewed evidence that it is of special salience to humans. Thus large fields of red light induce physiological arousal, and exacerbate the symptoms of certain neurological disorders. Red vision is most resistant to loss when colour vision is impaired by central brain lesions. Red is seen as a heavy, powerful, and active colour, and 'red' is usually the first colour word used by children.

These are dialectical relations that occur on an evolutionary time-scale; but individual development also involves dialectical relations between the individual and the environment. We have seen that red has special salience for humans. In the West we are taught that red is aversive, and so we use it as a danger signal; the use of red as a danger signal confirms that red is aversive to successive generations. In China, by contrast, red is similarly salient but has mostly positive connotations (Schafer 1967), and those positive connotations are confirmed to successive generations by its use for banners on ceremonial occasions and on New Year greeting cards.

Even in the West, red is not universally aversive. The good feelings of Christmas-time are very much tied up with the warmth of the colour red, and robin redbreasts figure conspicuously on our Christmas cards. Historical factors in our culture enter in here. My Oxford supervisor,

David Lack (1950), showed that the presence of robins on Christmas cards was related to the fact that the early postmen wore red waistcoats, and were known as 'Robin Postmen'; red post-boxes and red mail vans may well stem from the same source, marking out the territory of the Royal Mail. One might even go a step further and speculate as to whether the disappearance of red telephone boxes was a consequence not solely of practical considerations or of capitalist motivations, but of a desire to mark out a distinct empire for British Telecom. If so, we can trace the changes to the tendencies of individuals to seek to control their environments, or at least to build empires that they can feel that they control. This illustrates one theme of this chapter – the objects around us have their forms in part because of basic human propensities (the salience to us of the colour red and the collective assertiveness of individuals in the Post Office and British Telecom) and in part on aspects of history and fashion.

So far, I have presented evidence that the perception of colour constancy and the categorical perception of colours are independent of language, representing 'Relatively Stable human Characteristics', and I have suggested that they are adaptations to features of the physical world. In general we can say that our perceptual world, the environment in which we live, is influenced by our genes; but also that our genes have been selected in accordance with the structure of the physical world in relation to our way of life, or at least in relation to that of our remote ancestors. It suits bats and whales to be sensitive to the range of sound frequencies to which they are sensitive – bats for echo-location and whales for long-distance communication. Beyond that, I am suggesting that the meaning of many, possibly most, of the features of the world that we do perceive are established in a dialogue between the individual and the environment during development, with culture, itself determined in part by historical factors but also in part by biological ones, playing an important part in determining the course of that dialogue.

I must emphasize that I am not setting up a dichotomy between instinct and learning, between nature and nurture. Just the opposite. Because individual development involves a continuous transaction between what we are and what we experience, it is heuristically convenient to pick out what I shall call 'Relatively Stable Characteristics' – that is, characteristics of structure or behaviour that are basically similar in all individuals, or at least in all members of an age/sex class growing up under similar experiential conditions, though differing in detail between individuals (Hinde 1991). Thus we all have noses – noses are a Relatively Stable human Characteristic

– but all noses are different, so it is only a *relatively* stable one. For the most part, however, I shall be concerned here with relatively stable psychological characteristics of perception, motivation, and propensities to learn, characteristics that we all share, though not all to exactly the same extent.

THE ENVIRONMENT THAT WE DO EXPERIENCE

In living, we create the environment in which we live. By living, we influence that environment. How it affects us, what we create, and how we influence it depends in part on what we are. We seek out, create, or try to create an environment that suits what we are. We shall not fully understand the impact of the environment on our behaviour, or of our behaviour on the environment, unless we come to terms with the dialectical relations between them. This requires us both to pay attention to these basic and relatively stable psychological characteristics of mankind, and to their interaction with the experienced environment, including individual experience in the particular socio-cultural climate in which we live.

Before proceeding, it may be as well to illustrate the importance of considering this interaction with a simple example. Consider teddy bears, an artefact that humans create and which form a not unimportant part of the environment of many children. Why are they so popular as children's toys? Of course they are cuddly, and it is in the nature of children (indeed it is a Relatively Stable Characteristic of children) that they like to cuddle and be cuddled. The very property of cuddliness depends on the pleasure that humans have in cuddling. But why are teddy bears more popular than cuddly monkeys or cuddly dogs?

Teddy bears appeared around the turn of the century and were so called because of an association with the bear-hunting Theodore Roosevelt. The original teddy bears were very bear-like, with low foreheads and long snouts, and were often muzzled. But over the years they have undergone a change in physiognomy. By measuring the teddy bears in an exhibition in the Cambridge Folk Museum some years ago we were able to show that, on average, their foreheads have become higher and their snouts have become shorter. Of course, in any one year there is considerable variation, but the changes in the museum specimens between the beginning of the century and the early eighties were highly significant statistically.

Clearly, there is a problem here. Why did the teddy bear evolve?

A reasonable suggestion is that it is because teddy bears are bought by parents, and the characteristics have changed in a direction that is appealing to parents. Fifty years ago Konrad Lorenz (1950) suggested that the Disney animals were aberrant from their natural counterparts because they have features that elicit parental behaviour from humans, such as a high forehead and a short snout. These speculations have been confirmed experimentally (for example Fullard and Reiling 1976). Responsiveness to such stimuli is a Relatively Stable Characteristic. Apparently those bears that were most successful in attracting parents and thus in leaving the shop shelves each Christmas were more strongly represented there the next year (Hinde and Barden 1985).

Of course it is not quite so simple as that. There is considerable variation, implying to the evolutionary biologist that selection is not very severe. There are aberrant species, including the Pooh Bear, known as a bear of little brain, which has a low forehead. Fashion also plays a part, so that there may sometimes be a merit in being different. Perhaps that explains the curious trend towards longer snouts in the last few years, a trend that I predict will be short-lived (Hinde and Barden 1985).

There is a lesson here which I believe to be important. The characteristics of teddy bears depend in part on properties of the physical world – the texture of the furry surface. They depend in part on Relatively Stable human Characteristics (our propensities to be attracted to certain physiognomies and textures) and partly on history, on their origins, and the fashion that was then created for them by entrepreneurs.

The same is true of virtually all artefacts. Cars, for instance, must cope with air resistance, friction, and so on; they must fit round human bodies and satisfy human desiderata; they evolved from the horse and carriage and are subject to fashion. Rawson (1992), writing about an entirely different culture, independently came to a similar conclusion, namely that the particular form of decoration used on early Chinese bronzes depended on the properties of the material from which it was made; on properties of human perception; and on the history of and fashions in the type of artefact in question. In a sense, all the features of an artefact depend in the long run on human nature – our history and fashions were created by human activities, and the aspects of the physical world that we can perceive and that are important to us have been affected by the nature of humankind. It is important to remember that we create our culture, including the fashions that influence our behaviour. Culture is too often assumed to be something 'out there' that impinges on us. That is a false view. Culture is continually created, maintained, and changed by human activity, but the influences are two-way. We shape our environment and our culture, but they also shape us.

The most important part of our environment is our social environment: our personality is shaped by how we perceive others to perceive us, and in the short term we behave differently according to where we are and whom we are with. Yet at the same time we each have a sense of continuing self, each of us feels that he or she remains the same person although behaving differently. We resist the feeling of being chameleons because otherwise we should feel that we had no control over our lives or our environment. We do this in part by creating our own environment, an environment that suits us. Consider how we interact with others in social situations:

(a) So far as possible, we select those with whom to interact, making the selections in accordance with who and what we are.
(b) We behave differently according to whom we are with; tending to conceal from ourselves the changes in our own behaviour, saying 'I'm not really like that', or 'I'm only doing this because of that', creating for ourselves the impression that the chameleon-like behaviour indicates control.
(c) In interacting with others we not only choose a role for ourselves, but base the selection on how we perceive the person with whom we are interacting, and on how we perceive that that person perceives us, and on whether we feel that that person perceives us as we really are (for example Goffman 1959).
(d) Finally, we behave in a way that we believe will induce the other person to behave in a way that will enable us to behave as the sort of person we would like to be, and in a way that we perceive to be culturally appropriate. We may induce friendly behaviour by behaving in a friendly fashion, or submissive behaviour by being domineering.

Further, remember that all this applies equally to the person with whom one is interacting. He or she is also manipulating (and I use that word without any negative connotation) you. It will be apparent that our social behaviour involves complex negotiations with our social environment – a dialectic which brings a new social truth (for example Backman 1988; Hinde 1997). And we do not always recognize this because we need to feel that we have some control over our social environment – that we have a continuing self despite the slings and arrows that we encounter. Our relation with our social environment is like that with our physical environment in that we both affect and are affected by it.

But our present focus is on the physical environment, though I shall argue later that how that affects us may depend on the social environment. This too influences our behaviour from moment to moment in ways that depend on what we are. The posture one adopts when sitting in a chair is due in part to the shape of the chair, and the chair has the shape that it has in part because it was made to accommodate the shape and weight of the human body (as well as for reasons of history and fashion, and the sort of material of which it is made).

Let us consider an extreme case. In the early days of the space age, many experiments were performed to assess human performance under conditions of marked sensory deprivation. People were confined, floating in almost featureless, anechoic chambers. The results were devastating. Many of these people could not stand it for more than a few hours and showed marked disturbances in behaviour when released. The pictures we see nowadays of happy space people floating round their capsules implies that they do in practice get enough input, but this was certainly seen as a problem initially. We need variegated sensory input, and some aspects of our behaviour serve to satisfy that need.

Sensory input on its own is not enough. We also need to feel that we have some control over our environment. Some of the most effective techniques for indoctrination and brain-washing, used for instance in some conflicts on prisoners-of-war, consist of depriving the subject of any control over what happens to him or her. Indeed this is, or was, part of the normal procedure for inducting recruits into the armed services. And many aspects of our personal behaviour and relationships can be traced to the need to feel in control.

Given the need for sensory input and for some control over it, can we make any further generalizations about the sorts of objects or environments that we prefer, and the effects that they have on us? It is important here not to take too limited a viewpoint, and I shall emphasize shortly how history and fashion, mediated by individual experience, influence what we like to perceive. But first let us ask whether there is any evidence for pan-cultural basic human propensities to prefer and respond appropriately to some sorts of input and environment, rather than others. At one level, of course there is. We pick out figure from ground, tend to see figures as complete, are sensitive to symmetry and other sorts of patterning, and so on. We need not be concerned with whether or not such abilities depend on experience: the abilities of young babies show that experience of a general sort such that virtually all babies encounter is adequate (for example Mehler and Dupoux 1994). A large body of evidence shows that newborns respond preferentially to the stimuli provided by the breast, and appropriately to loud noises and loss of support.

But more relevant in the present context is some evidence concerning responsiveness to characteristics of the environment as a whole. A convenient starting point is E.O. Wilson's (1993) biophilia hypothesis. He suggested that we have an affinity for the natural world, an affinity which, he suggests, results from the action of natural selection earlier in human history. In speaking of an affinity he implied a predisposition to respond positively, or to learn to respond positively, to other human beings and to aspects of the natural world. Predisposition implies a Relatively Stable Characteristic, one likely to have arisen under the influence of natural selection. There is, in fact, evidence in harmony with the view that humans respond positively to certain types of natural landscapes, and with aversion to certain natural stimuli.

Four views about basic human responses to landscapes, not incompatible with each other, can be distinguished. The first simply emphasizes the preference for natural over artificial landscapes. Orians and Heerwagen (1992) and Ulrich (1993) have reviewed a number of studies in which people are asked to rank-order pictures of landscapes. Natural landscapes are generally preferred over urban views. Artificial features, such as power-lines, tend to detract from natural landscapes, and natural features to enhance urban views. There is also considerable evidence that natural landscapes are special in that they have stress-reducing properties. One study showed that recreation in wilderness areas brought relief from stress, when a number of other possible explanations were controlled. Others demonstrate that slides or pictures of urban scenes reduce stress or induce a positive emotional tone less than those of natural scenes. Dental, surgical, and psychiatric patients are helped by pictures of scenic views, and prisoners able to see farmlands and forests are less likely to be sick than those whose cells permit a view only of the prison yard (Heerwagen and Orians 1993; Ulrich 1993).

Further evidence comes from an extensive study in the New York area, covering a range of social classes and mostly in urban areas. Halle (1993) found that landscapes were either the first or second most common topic of pictures in three quarters of the houses sampled, and constituted a third of all pictures displayed. The only exception was an area of great scenic beauty on Long Island, where one quarter of the houses had no landscapes at all. Three hundred and forty-seven out of three hundred and forty-nine of the landscapes were of calm and peaceful scenes, the two exceptions being accountable for in terms of personal associations for the owners. Furthermore over half of the respondents who liked abstract art said that they did so because it brought to mind images of landscapes. (These, we should note, are the landscapes we are destroying.)

A second view, slightly more specific, suggests that, amongst landscapes, it is savannah that is most preferred. Savannah environments were favourable for early humans, and much early human evolution took place there. They were better suited to organisms with an upright posture, offered abundant food and more opportunities for surveillance or escape and lower probabilities of hidden dangers. It has therefore been suggested that humans have an evolved tendency to prefer, or to learn to prefer, environments with savannah- or park-like properties, and to avoid both absolutely open spaces on the one hand and dense forests on the other. It is also suggested that landscapes having water should be especially preferred.

There is some evidence in favour of this suggestion. A number of studies show that European, American, and Asian adults respond more positively to savannah-type environments than to others. One study used six age-groups of people, none of whom had ever been in tropical savannah, and assessed their preferences for a variety of biomes. The youngest (8 years old) showed a clear preference for savannah. From age 15 years onwards savannah, deciduous, and coniferous forest were liked equally well. The authors suggest that the early predisposition to prefer savannah was modified subsequently by the individual's experiences (Balling and Falk 1982).

A third view, closely related to the last, focuses on the specific features of an environment likely to make it attractive to humans. These include many of the features of savannah, including openness but with protective cover and the availability of refuges and situations from which one could see without being seen and assess risks. Trees with moderately dense canopies and whose trunks divided near the ground seem to be especially attractive: such trees are characteristic of the more fertile parts of the savannah (Orians and Heerwagen 1992). In harmony with this view, Appleton (1975), surveying a range of English landscape paintings, suggested that 'Prospect', having a good view of the landscape, and 'Refuge', having a good place to hide and from which one could see without being seen, were determinants of preference.

Orians and Heerwagen (1992) emphasize that in natural circumstances decisions about a particular environment are likely to be taken in stages, and to take into account both immediate advantages and prospects for the future. For example flowers usually presage fruit and seeds, and thus indicate an environment likely to provide plentiful food in the near future. They speculate that this could explain the attractiveness of flowers. (Could it also explain the lesser importance of flowers in Africa than elsewhere (Goody 1993) on the grounds that seasonality is less marked in tropical regions?)

The fourth approach (Kaplan 1992), rather different and more abstract, supposes that human responses to landscapes depend on human needs to understand and to explore, and on how much cognitive processing is necessary before inferences can be drawn. In an extensive series of experiments 'Mystery' – the promise of more information if one could enter further into the scene – was the strongest predictor of preference. Absence of immediately apparent characteristics, requiring minimal inference, was associated with low preference. Absence of complexity means that there is little to look at, absence of coherence makes a scene difficult to understand.

There is a current controversy in studies (or speculations) about human evolution. Are we to postulate that natural selection has endowed us with a large number of Relatively Stable Characteristics, each controlling a particular aspect of perception or behaviour (for example Buss 1994), or should we try to explain our behaviour in terms of relatively few propensities which, through their mutual interactions and through interactions with the socio-cultural structure consequent upon them, produce the complexities of our lives (Hinde 1987)? To be specific, is the attractiveness of savannah due to responsiveness to a number of specific features, like the shapes of trees, or to a few basic propensities, like the need for food and the need to survey safely and thus to control the environment? Again, if 'mystery' is an important element in our response to landscapes, is that 'mystery' specific to that response, or is it related to the motivation to unravel mysteries inherent in research, and even to the mystery said to play a part in the allure of sex? Probably the truth lies somewhere between these two possibilities. In any case, the issue here is that the evidence, although drawn from relatively few societies and over a limited span of time, is compatible with the view that humans have predispositions which lead them to prefer certain types of landscapes over others. There is also evidence (reviewed Ulrich 1993) that humans are predisposed to learn to respond, usually with aversion, to certain natural objects, such as snakes, as discussed below.

Even given certain predispositions, we must expect individual experience to have moulded preferences: the forest dwelling pygmies, the Bedouin, the Inuit, inhabit and presumably prefer habitats other than savannah. And even within our own society there have been changes with time. Sir Keith Thomas (1983) has traced general trends in landscape preferences in Britain from the Renaissance, when the city was synonymous with civility and the country with boorishness, to the late eighteenth century, when the countryside was seen as more beautiful than the town. The latter view was reinforced as towns became increasingly smelly, dirty, and crowded in the Industrial Revolution.

Nevertheless, Thomas (1983) states that, from time immemorial, most people have been attracted to large views and open prospects, and he acknowledges a continuing degree of ambivalence. Wordsworth was struck by the beauty of Westminster Bridge as well as that of the daffodils. Halle's study of the artefacts in New York houses contains a beautiful example of this ambivalence and of changing attitudes to the countryside. Only 11% of the landscapes of contemporary America contained human figures, whereas 71% portraying past scenes or non-industrialized countries did. On the basis of interview material, Halle ascribes this to the modern view that man desecrates the natural world. In general, we must not let the nature/nurture dichotomy intrude on our thinking: both aspects of human nature (Relatively Stable Characteristics) and experience influence what we like.

Goody (1996) has emphasized that ambivalence about man's role in nature is not limited to the modern industrialized world, arguing that the 'so-called animist or nature-worshipping religions . . . can be seen . . . as part of a dialogue with and about nature' in which humans must make recompense for the damage they had inflicted on nature. Goody suggests, for instance, that the harvest sacrifice is to be seen not just as thanksgiving but also as an apology for the killing of the corn spirit.

Other material reviewed by Thomas might suggest that fashions for landscapes may be influenced also by the personalities of the beholders. In the eighteenth century there was a vogue for pictures of mountain scenery, related to the mystique of mountain-climbing. We now know that young men show an increase in risk-taking behaviour which peaks in the early twenties and then declines. It would be of the greatest interest to know whether the fashion was instigated by young men or by those who wished that they were still young men. But it must be remembered that landscapes may be attractive for a great variety of reasons; and liking is not the same as fascination.

Furthermore the routes by which we invest landscapes with

meanings are diverse. The Rock of Gibraltar became a symbol of the power and permanence of the British Empire and, thanks to Dunkirk and Vera Lynn, the white cliffs of Dover became a bulwark against the enemy. Similarly Mount Fuji is a symbol of Japan. This theme has recently been elaborated by Simon Schama (1995), arguing that 'nature' is created by human minds and cultural significance is ascribed to diverse environments in diverse ways. Thus woodland and forest have stood for liberty under the greenwood in England, for transcendental connections with the Almighty in the USA, and for the fight for national independence in Poland.

The important issue here, too often forgotten both by those who emphasize biological influences and by those who describe how fashion influences preferences, is that Relatively Stable human Characteristics, cultural influences including fashion, and personal experience, may all operate in our taste for landscapes and pictures, just as they seem to do in the case of the teddy bear.

Similar principles operate for aversions, a paradigmatic example being provided by responses to snakes. Snakes have a special salience for children from a year or two in age. Whether or not they are afraid, and how afraid, depends on the behaviour of their mother or other caregiver: if the latter shows fear, the child is likely also to show fear and to be afraid of snakes thereafter. Whereas the evidence here comes largely from clinical case studies, an elegant series of experiments by Mineka (1987) has demonstrated that similar principles apply to rhesus monkeys:

(a) Wild-reared rhesus monkeys nearly always show fear responses to a snake.
(b) Laboratory-reared rhesus monkeys show little or no fear of snakes.
(c) Laboratory-reared rhesus monkeys, shown a video-tape of a wild-reared monkey giving a fear response to a snake, become afraid of snakes thereafter.
(d) Laboratory-reared rhesus monkeys shown a precisely similar video-tape, but with the wild-reared monkey appearing to give a fear response to a flower, do not become afraid of either snakes or flowers.

Thus in both humans and rhesus monkeys the fear response to snakes appears to depend both on a relatively stable biological predisposition and on learning from experience. But in humans there is another important issue. Snakes play a very important part in our cultural imagery. In the West they symbolize evil. The Rubens paintings of lost souls going down to Hell with snakes gnawing at their genitals must have made people feel that it was very desirable not to go to Hell. In other cultures, snakes carry other messages, symbolizing royalty, health, and power. The response to snakes, like that to the colour red, depends both on an initial predisposition and on experience in the culture (Hinde and Rawson 1995). The importance of basic predispositions is demonstrated by the fact that all the true human phobias – for example of falling, darkness, wide open spaces, or being shut in – concern issues that would have been of danger to our remote ancestors in our environment of evolutionary adaptedness (Marks 1987). The importance of cultural influences is shown by the fact that the salience of snakes is interpreted in a positive fashion in some cultures – especially, for instance, in southern India. Although, as elsewhere, snakes are generally regarded with aversion, they are seen also as a source of power. The primal snake, the so-called 'snake without end', acts as the bed-rest of Vishnu, and is seen as the source of universal energy and cohesion. The association of snake forms with the root of universal power gives them a special significance. One may speculate as to whether their use as symbols of power or protection does not have an ultimate basis in fear (see also Mundkur 1983). Other cultural symbols, like the eagle, the lion, or the springbok, probably depend wholly or primarily on characteristics that we have learned to associate with them.

In discussing our preferences for and aversions to environments and to natural objects in our environment, it may seem that I have strayed too far from my brief to discuss the effects of the environment on behaviour. But we choose the environments, the depictions, and other artefacts that we like to have in our houses because they make us feel something that we want to feel. Halle found that the owners of the pictures said that they liked the peaceful pictures because they were calm, restful, and so on, and because they liked to imagine being there – and sometimes because the scene had special associations for them. We design churches to foster reverence, libraries to encourage study, hospitals to facilitate the treatment of patients, conference halls to enable discussion. In general, the environments we prefer or try to create are consequences of what we are and of what we desire.

Human artefacts form a major part of our environment, and we create them to satisfy basic human propensities. We surround ourselves with devices to obtain water, keep ourselves warm or cool, comfortable, entertained, and so on; but every one of these artefacts is there because of some aspect of human nature, and also because of some aspect of the physical world. The chromium taps on the bath are descended from more glorious Victorian brass ones. However let us consider some other reasons for which we choose the objects with which we surround ourselves.

A high proportion of the objects in peoples' houses are there because of their associations. Often the association is with recent events or current loved ones, for instance through the depictions of people in family photographs. Halle's New York study again provides an interesting example. Items of so-called 'primitive art' were present in 40% of upper class white homes and of middle class Afro-American homes, but were almost absent from working class white homes. The Afro-Americans also had classical religious depictions with black figures, artistic representations of Afro-Americans in everyday scenes, Afro-American political figures and entertainers, and all had family photographs. For them the 'primitive' art expressed and affirmed their current lives and their culture. By contrast, with the Whites it symbolized attitudes to Afro-Americans, confirming boundaries. The working class and some middle class White respondents disliked 'primitive' art because they saw it as fearsome. Democrats saw it as co-equal with western art: Republicans did not.

Sometimes the association takes the form of what may be called sympathetic magic. Rozin and Nemeroff (1990) have pointed out that, although we usually regard sympathetic magic as characteristic of people less sophisticated than ourselves, in fact it plays a not unimportant role in our lives. They mention three sorts, of which two may be mentioned in the present context. The Principle of Similarity refers to the equation of image with object, as when we feel that positive or negative consequences will follow from contact with a replica or image, or that action on a replica will affect the original. We like to possess likenesses of those we love. And, while few of us would stick a pin in a photograph of a rival, we might enjoy tearing it up. On the principle that it is better to respond to something that looks like a tiger as though it was a tiger, the principle of similarity may be more than a simple matter of stimulus generalization and have been selected for in our evolutionary history.

The Principle of Contagion concerns beliefs that things that have once been in contact may influence or change each other for a considerable time thereafter. Thus we treasure a gift from a loved one, even though the object itself is not to our liking, as if it contained an essence imparted by the donor; and there is always a market for the possessions of the famous. The habit of exchanging clothes or rings between romantically involved couples may involve the literal or metaphorical sharing of an essence: the dangers inherent in eating a sugar lump on which a fly has settled suggest that this principle also may have a biological basis.

The associations may also be with our recent past, important perhaps because they remind us of happy times. We collect objects associated with happy times in part because it is part of our nature to like to be reminded of happy times. They may also give us a sense of continuity in other ways. Gravestones and memorials give us a sense of contact with loved ones, confirming the narrative we have woven about the role they have played in our lives and ameliorating the sense of loss. Bereavement in war is harder to bear if the dead one has 'no known grave', and next-of-kin often undertake arduous journeys to visit a military cemetery.

The association may go back much further, the object serving to establish a much longer narrative involving contact with roots. In affirming their culture, the 'primitive art' in the homes of the Afro-Americans studied by Halle also established a contact with their ancestral homeland and with past generations. National monuments likewise serve to maintain continuity with past generations. In the even longer term, Rawson (1993) has argued that the revival of Shang and Zhou bronze types 2000 years later in the Sung and Yuan dynasties created a possibly artificial continuity with the past which may have been important for political reasons, for the Chinese pride themselves that they can trace their history continuously over at least 4000 years.

The importance of ancestral homes and the fascination in seeing where parents grew up must also depend on a desire to maintain contact with the past, to extend even further back the narrative of our existence. Hence also the value placed on the age of objects.

While monuments may encourage current generations to emulate the achievements of past ones, sometimes their function is more sinister – to remind people of the suffering of past generations in the hands of enemies or oppressors and thus perpetuate a diachronic transference of pain and a desire for revenge. An example of this is to be found in the two Museums of National Struggle in the Greek and Turkish sections of the divided city of Nicosia. They present history in radically opposing ways by subsuming the history of Cyprus into that of Greece on the one hand and Turkey on the other. Each perceives the other as an historic enemy, emphasizing their own heroes in this 'righteous' struggle (Papadakis 1995).

Going even farther back, the origin myths, found in every culture and providing perceived understanding of where we came from, are part of the stock-in-trade of anthropologists. It seems that we need to construct a coherent narrative of our lives, including our origins, so that we have a sense of being in control. It is tempting to ask whether the desire to know what happens after death, to postulate other worlds that we shall inhabit in other forms, can also be understood in terms of a need to feel in control. Perhaps we feel that the longer our roots, the further our branches will extend? Can both be ascribed to a need to perceive continuity in our lives, and that in turn to the need to feel in control?

Artefacts also play an important part in conveying social relations and status. Coming from one society to another, both of which love to display their multi-coloured gowns, I do not need to emphasize this point. But it is worthwhile pointing out that at least some status symbols exaggerate characteristics to which we are predisposed to respond. Thus many military helmets exaggerate height, and the epaulets increase the apparent width of the shoulders (Eibl Eibesfeldt 1970, 1975; Koenig 1968). Ornament is used in a similar way to accentuate sexuality – lipstick makes lips redder, bras accentuate the bosom, and corsets produce a body form more closely approximating to a desirable waist/hip ratio.

Status is also often emphasized by the possession and display of exotic objects. Their value stems in the first instance from the fact that, coming from a great distance, they were necessarily costly and few could have access to them (Rawson 1992/3). In the case of oriental carpets, Spooner (1986) has traced the way in which, as technology caused distances to shrink, exclusivity gave way to authenticity. As reproduction became more possible, the dialogue between source and consumers became more direct, and more consumers could compete for possession. The only way to preserve the specialness of the carpets was then to complicate the criteria of authenticity. Thus the basic desire for status has led to a dialogue between source, consumer, replicator, and connoisseur.

What is interesting here is that status symbols may be so gross that they are a real handicap to their owners. People buy cars, jewelry, and houses that are way beyond their means, primarily to show off. There are analogies here in animals – some species have evolved accoutrements, such as the peacock's tail, which, though disadvantageous to personal survival, are attractive to the opposite sex (Zahavi 1977). Objects also contribute to social integration. Flags, uniforms, and medals help to fuel patriotism and nationalism – playing on basic human propensities – seeking for status, cooperation with in-group members, and the denigration of the out-group, and so on.

In all these cases, artefacts are either helping us to feel what we are predisposed to want to feel or helping to make others respond to us as we would want them to respond. In other words, they are assisting the realization of basic human propensities.

Now we may return to the issue with which we started – that many of the effects that we have on the environment, and the artefacts that we create, even when they are in the short-term interests of some individuals, are to the long-term detriment of others. Thus the problem of the common land becomes more and more conspicuous in our society. It is not just a question of 'If I do not put all the sheep I can onto the common land, someone else will', or 'Just one more sheep won't make any difference', because the issues have escalated. The behaviour of individuals still matters, but we also have to contend with mammoth organizations arguing that the oceans are so large that the little bit of pollution that they add will not make any difference, or that the sky is so large it can easily absorb a little bit of sulphur from their chimneys. Such rationalizations have complex bases. Each organization has its own ethos, its own goals. A successful organization comes to resemble a psychological group in that its members come to identify with it, to elaborate and share common goals, and to consider their organization better than others and its members better than outsiders. The goals may be related to satisfying behavioural propensities in other individuals – providing fuel to warm our houses, food for us to eat, providing the so-called necessities of life (though they also create new needs which crave satisfaction), but they may also produce consequences undesirable both for those inside and for those outside the organization, and even for future generations. The undesirable consequences are pushed to one side because, within each organization, group dynamics operate to produce group loyalty. Individuals work to achieve the group's goals and vie with each other in advancing the group's interests. Individuals who are successful in promoting the organization's goals gain kudos. In so far as they are conscious of the destruction that results from their efforts, they may conceal it from themselves: the phrase 'I'm not really like that' comes in useful here too. Personal ambition and assertiveness are behind the collective greed, and the collective greed can produce changes in the environment that are to the detriment of the many.

To be fair, it is not only greed and ambition, but also ignorance. We have to be told that, if enough of us satisfy our desire to climb mountains, to picnic in the dunes, or to pick the meadow flowers, we shall spoil them for others At the organization level, knowledge of the damage being caused is too easily countered by rationalizations. One such is the belief that unlimited economic growth is desirable and possible, another is mistaken faith in the ultimate availability of technical fixes for all our woes – a belief unthinkingly fostered by some scientists, perhaps as a consequence of their desire for status or to feel in control; another is the view that the earth's resources are unlimited (Arrow *et al.* 1995; Smith *et al.* 1994).

I have argued that what we do to the environment, and thus what we cause the environment to do to us, is a product of our nature. While what the individual does to the environment may be good for him or her, it is not necessarily good for others. But that certainly does not mean that we are caught in an inexorable spiral. If we understand what is happening we can break out. At both the individual and the organizational levels education and discussion are necessary. That is why this lecture series sponsored by Linacre College is so important. In considering the effects of the environment on our behaviour, it is essential to remember that the effects are two-way – we create the environment that impinges on us. While the work of chemists, physicists, and biologists is essential for understanding what we are doing, we must look to psychology, sociology, and economics to understand why. It *is* possible to change group goals and group values, and it is essential that we start to do so.

In conclusion, accepting that the environment affects our behaviour, I have tried to argue that we must see the environment not as a given 'out there', but as our own creation in both metaphorical and literal terms. We create our environment by exercising choice as to where we should live or where we should go on our holidays, and what we choose is a function of what we are. We create our environment by surrounding ourselves with artefacts which satisfy our natures in diverse ways. We ascribe meanings to scenes, to natural features, to artefacts, which in turn affect our behaviour in ways that we deem to be desirable. By influencing the environment in ways that satisfy their own propensities, individuals may act to the detriment of the many. However, such an outcome is not inevitable. If we examine our goals and values, it should be possible to find a way out before it is too late.

REFERENCES

Appleton, J.(1975). *The experience of landscape*. Wiley, New York.
Arrow, K., Bolin, B., Costanza, R., Dasgupta, P., Folke, C., Holling, C.,

et al. (1995). Economic growth, carrying capacity, and the environment. *Science*, **268**, 520–1.
Backman, C.W. (1988). The self: a dialectical approach. *Advances in Experimental Social Psychology*, **21**, 229–60.
Balling, J.D. and Falk, J.H. (1982). Development of visual preference for natural environments. *Environment and Behaviour*, **14**, 5–28.
Berlin, B. and Kay, P. (1969). *Basic color terms*. California University Press, Berkeley, California.
Buss, D. (1994). *The evolution of desire*. Basic Books, New York.
Eibl Eibesfeldt, I. (1970, 1975). *Ethology*. Holt, Rinehart, and Winston, New York.
Fullard, W. and Reiling, A.M. (1976). An investigation of Lorenz's babyishness. *Child Development*, **47**, 1191–3.
Goffman, E. (1959). *The presentation of self in everyday life*. Doubleday Anchor, New York.
Goody, J. (1993). *The culture of flowers*. Cambridge University Press, Cambridge.
Goody, J. (1996). Man and the natural world: reflections on history and anthropology. *Environment and History*, **2**, 255–69.
Halle, D. (1993). *Inside culture*. Chicago University Press, Chicago.
Heerwagen, J.H. and Orians, G.H. (1993). Humans, habitats, and aesthetics. In *The biophilia hypothesis* (ed. S.R. Kellert and E O. Wilson), pp.138–72. Island Press, Washington, DC.
Hinde, R.A. (1987). *Individuals, relationships, and culture*. Cambridge University Press, Cambridge.
Hinde, R.A. (1991). A biologist looks at anthropology. *Man*, **26**, 583–608.
Hinde, R.A. (1997). *Relationships: a dialectical perspective*. Psychology Press, Hove.
Hinde, R.A. and Barden, L. (1985). The evolution of the teddy bear. *Animal Behaviour*, **23**, 1371–3.
Hinde, R.A. and Rawson, J. (1995). Ritual: signals and meanings. In *Rites et ritualisation* (ed. G. Thines and L. de Heusch), pp.21–36. Vrin, Paris.
Humphrey, N.K. (1983). *Consciousness regained*. Oxford University Press, Oxford.
Kaplan, S. (1992). Environmental preference in a knowledge-seeking, knowledge-using organism. In *The adapted mind* (ed. J. Barkow, L. Cosmides, and J. Tooby), pp.581–98. Oxford University Press, New York.
Koenig, O. (1968). Biologie der Uniform. *Naturwissenschaften Medizin.*, **5**, 3–19 and 40–50.
Lack, D. (1950). *Robin redbreast*. Clarendon Press, Oxford.
Lorenz, K. (1950). Ganzheit und Teil in der tierischen und menschlichen Gemeinschaft. *Studium Generale*, 1950, 3/9.
Maloney, L.T. and Wandell, B.A. (1986). Color constancy: a method for recovering surface spectral reflectance. *Journal of the Optical Society of America*, **3**, 29–33.
Marks, I.M. (1987). *Fears, phobias and rituals*. Oxford University Press, New York.

Mehler and Dupoux, E. (1994). *What infants know*. Blackwell, Cambridge, Mass.

Mineka, S. (1987). A primate model of phobic fears. In *Theoretical foundations of behaviour therapy* (ed. H. Eysenck and I. Martin). Plenum, New York.

Mundkur, B. (1983). *The cult of the serpent*. State University of New York Press, Albany.

Orians, G.H. and Heerwagen, J.H. (1992). Evolved responses to landscapes. In *The adapted mind* (ed. J. Barkow, L Cosmides, and J. Tooby), pp.555–80. Oxford University Press, New York.

Papadakis, Y. (1995). Nationalist imaginings of war in Cyprus. In *War: a cruel necessity?* (ed. R.A. Hinde and H. Watson), pp.54–67. Tauris, London.

Rawson, J. (1992). Late Shang bronze design: meaning and purpose. In *The problem of meaning in early Chinese ritual bronzes. Colloquies on Art and Archaelogy in Asia*, no.15 (ed. R. Whitfield), pp. 67–95. School of Oriental and African Studies, London.

Rawson, J. (1992/3). Contact between Southern China and Henan during the Shang period. *Transactions of the Oriental Ceramic Society*, **57**, 1–24.

Rawson, J. (1993). Ancient Chinese ritual bronzes: The evidence from tombs and hoards of the Shang and the Western Zhou periods. *Antiquity*, **67**, 805–23.

Rozin, P. and Nemeroff, C. (1990). The laws of sympathetic magic: a psychological analysis of similarity and contagion. In *Cultural psychology* (ed J. Stigler *et al.*). Cambridge University Press, Cambridge.

Schafer, E.H. (1967). *The vermilion bird*. University of California Press, Berkeley, California.

Schama, S. (1995). *Landscape and memory*. Harper Collins, London.

Shepard, R.N. (1992). The perceptual organisation of colours: an adaptation to regularities of the terrestrial world. In *The adapted mind* (ed. J. Barkow, L. Cosmides, and J. Tooby), pp. 495–533. Oxford University Press, New York.

Smith, P.B., Okoye, S.E., Wilde, J. de, and Deshingkar, P. (eds) (1994). *The world at the crossroads* (A report to the Pugwash Council). Earthscan, London.

Spooner, B. (1986). Weavers and dealers: the authenticity of an oriental carpet. In *The social life of things* (ed. A. Appadurai), pp.195–235. Cambridge University Press, Cambridge.

Thomas, K. (1983). *Man and the natural world*, p. 265. Allen Lane, London.

Ulrich, R.E. (1993). Biophilia, biophobia and natural landscapes. In *The biophilia hypothesis* (ed. S.R. Kellert and E.O. Wilson), pp.73–137. Island Press, Washington, DC.

Wilson, E.O. (1993) Biophilia and the conservation ethic. In *The biophilia hypothesis* (ed. S.R. Kellert and E.O. Wilson), pp.31–41. Island Press, Washington, DC.

Zahavi, A. (1977). The testing of a bond. *Animal Behaviour*, **25**, 246–7.

2
How the environment helps to build the brain

Colin Blakemore

Professor Colin Blakemore, PhD, ScD, FRS, studied medicine at the University of Cambridge and completed his PhD in Physiological Optics at the University of California, Berkeley, in 1968. After teaching at Cambridge for eleven years, he was appointed Waynflete Professor of Physiology at the University of Oxford in 1979 and has held the post ever since. Professor Blakemore is also Director of both the Medical Research Council and the McDonnell-Pew Centres for Cognitive Neuroscience in Oxford and is President of the British Association for the Advancement of Science for 1997–98. During 1995–96 he was Regents' Visiting Professor at the University of California, Davis. He was elected a Fellow of the Royal Society in 1992 and has been the recipient of numerous prizes and medals from medical and scientific institutions throughout the world. Professor Blakemore is a frequent broadcaster on radio (he delivered the BBC's Reith Lectures in 1976) and television; he presented a 13-part television series on the mind and brain, The Mind Machine, *in 1988. In addition to his many academic publications, he has written several books on the brain for the general public, including* Mechanics of the Mind, *for which he was awarded the Phi Beta Kappa Award in Science, and* The Mind Machine. *He is one of Britain's most influential communicators of science.*

The problem of understanding the mature human brain, with its 100 000 million or more nerve cells, its 3 kilometres of nerve fibres, and its astronomical numbers of precisely organized connections, pales into virtual insignificance compared with the challenge of comprehending the development of this extraordinary organ. Just consider the magnitude of this puzzle. We suspect that the human chromosomes contain perhaps 100 000 genes. Each gene is a simple instruction – a formula for the construction of a protein. Some of these proteins contribute to the structure of cells throughout the body; some act as enzymes, catalysing reactions that can create other sorts of molecules; some are signalling factors, including hormones and growth factors, and the transmitter substances that convey messages across the tiny gaps between nerve cells.

Although more than 25 000 of our genes are involved in making the nervous system, this number of instructions seems pitifully small compared with the magnitude of the task to be specified. The blueprint of the brain, contained in the genes, does not seem to match up to the complexity of the object to be built – like constructing the Taj Mahal from a sketch on the back of an envelope.

There is another, related problem. Every cell in the body (except the eggs or sperm) contains a full set of all the chromosomes, and all their genes. The differences in appearance and function between, say, a liver cell and a kidney cell are due to differences in the *expression* of those genes – the regulation of the timing and magnitude of production of proteins. Within the developing embryo, interactions between all the proliferating cells steer them along particular lineages of *fate*, each type of cell turning on a subset of all the genes and hence becoming specialized according to the genes that are expressed. This divergence of form and function among the population of cells in the body, according to the genes that are expressed – a process termed *differentiation* – is one of the fundamental problems in biology.

Nowhere is differentiation more striking and more puzzling than in the brain. While the liver and the kidney each consist of perhaps half a dozen different sorts of cells, the number of distinct cell types in the brain is orders of magnitude larger. First, there are two broad classes – the *glia* (meaning 'glue') cells, which serve a variety of supportive roles but do not transmit impulses; and the *neurons* themselves, most of which have long processes called *axons*, along which electrical impulses are conducted. But the population of neurons can be subdivided into at least hundreds of types. The shapes and sizes of their cell bodies vary enormously. So too do the patterns of their *dendrites* (the spidery processes that stick out from the cell body, on which large numbers of incoming axons terminate), and the lengths and distributions of their axons.

The junction between the terminal of an axon and the dendrite or cell body of the next neuron in the chain is called a *synapse* and the signal is communicated across the tiny gap between the two by a chemical *transmitter substance* secreted by the axon terminal whenever impulses arrive along that axon. The nature of the transmitter substance (or substances) secreted varies from neuron to neuron. Some transmitters are excitatory, triggering changes in the next cell that tend to make it fire impulses; others are inhibitory, decreasing the probability of the next cell firing.

The most significant form of variation between different classes of neuron is in the exact pattern of incoming connections (which govern the functional properties of the neuron) and in the distribution of the outgoing signals to other cells (which determine the nerve cell's effects on the rest of the brain). It is the richness and precision of the interconnection of neurons that gives the brain its remarkable ability to process information, to form representations of the outside world, and to generate adaptive behaviour.

How can all the variety of neurons and their connections be produced by the action of such a limited repertoire of genetic instructions? The short answer is that it is not – at least not in its entirety. The development of the brains of higher mammals involves an interplay between the slavish instantiation of innate programs and subtle influences from the environment of the animal, steering the course of development and hence imposing order and constraint on the structure and function of the brain.

THE CEREBRAL CORTEX: THE APOTHEOSIS OF DIFFERENTIATION

Nowhere in the body is differentiation more spectacular than in the cerebral cortex, the vast folded mantle that is the most prominent feature of the mammalian brain. This sheet of tissue, about 3 mm thick and 1600 cm^2 in area in a human being, is undoubtedly responsible for the 'highest' and computationally most complex capacities of the nervous system – sensory perception, conscious memory, thought, voluntary movement, language, and consciousness itself.

Superficially, the entire cerebral cortex appears boringly consistent in its structural organization. Slices of cortex viewed through a microscope look, to an untutored eye, remarkably similar, wherever they come from in the cerebral hemispheres. Six layers of fibres and cells can be discerned. The uppermost layer (layer 1) is mainly a plexus (or network) of fibres with very few cells. Of the remaining five layers the middle one, layer 4, is distinctive because of its closely-packed, small cells. These tiny neurons receive most of the input to the cortex, delivered along axons that arise from a huge structure, the *thalamus*, lying in the middle of the brain, below the cerebral cortex. The thalamus itself receives fibres from the major sense organs, as well as from other groups of cells, in the stem of the brain, concerned with the control of movement.

The axons of the small cells of layer 4 do not, in general, leave the grey matter of the cortex: instead they ramify over quite short distances, delivering signals to the cells in the other layers, above and below. It is a fair approximation to say that the cells in the upper layers, layers 2 and 3, mainly distribute their axons to other regions of the cortex, both nearby and over long distances, while those in layers 5 and 6 mostly send theirs downwards throughout the lower part of the brain and into the spinal cord. In general, neurons with long and/or thick fibres have larger cell bodies: hence the cells of the lower layers of the cortex tend to be particularly large; and the largest of them, called *pyramidal cells*, have prominent main dendrites running straight up towards the surface of the cortex.

Intermingled within the circuits of processing neurons are large numbers of inhibitory cells, of various distinctive forms, whose axons are distributed amongst the neighbouring excitatory cells. Inhibition is an essential component in the function of the cortex: the terrible convulsions of an epileptic fit are essentially due to a failure of inhibition, leading to spasms of uncontrolled excitation rushing through the circuits of the cortex.

Against this background of apparent uniformity of organization is a remarkable pattern of local specialization. In the fifth century BC, Alcmaeon of Croton, one of the founders of neurology, proposed that humans alone have the power of understanding; he considered sensation and thought (or understanding) to be separate processes (Rose 1994). Moreover, he suggested that each major sense has its own region in the brain fed by a nerve from the sense organ. This notion of separate, parallel processing of information has been thoroughly vindicated through subsequent studies of the cerebral cortex. The cortex is divided into distinct territories or fields, with more or less sharp boundaries between them.

After the discredited efforts of the phrenologists, from the end of the eighteenth century, to assign such characteristics as 'sublimity', 'benevolence', and 'amorousness' to discrete 'organs' across the surface of the brain, legitimate evidence for localization of function in the cerebral cortex grew rapidly and inexorably (see Clarke and O'Malley 1968; Blakemore 1977), mainly derived from the effects of brain damage in humans. Pierre-Paul Broca and Karl Wernicke postulated areas concerned with different aspects of language; and

John Hughlings Jackson not only described a region of cortex concerned with the control of motion but also suggested that this area is subdivided, with different segments devoted to the hand, the foot, the lips, and so on.

The doctrine of localization of function in the cortex has been fully confirmed through the many techniques now available to neuroscience. Electrical stimulation of the human cortex, behavioural studies of animals after cortical damage, the electrical recording of impulses in animals, and most recently, the imaging of activity in the brains of normal living human volunteers and clinical patients, have all supported the view that the cortex is a patchwork of distinct functional areas, each committed to the processing of a particular aspect of a sensory input or motor output, the storage of memories, or the planning of behaviour. The multitude of highly specific connections between these areas creates a number of parallel, but partly interconnected 'streams' of

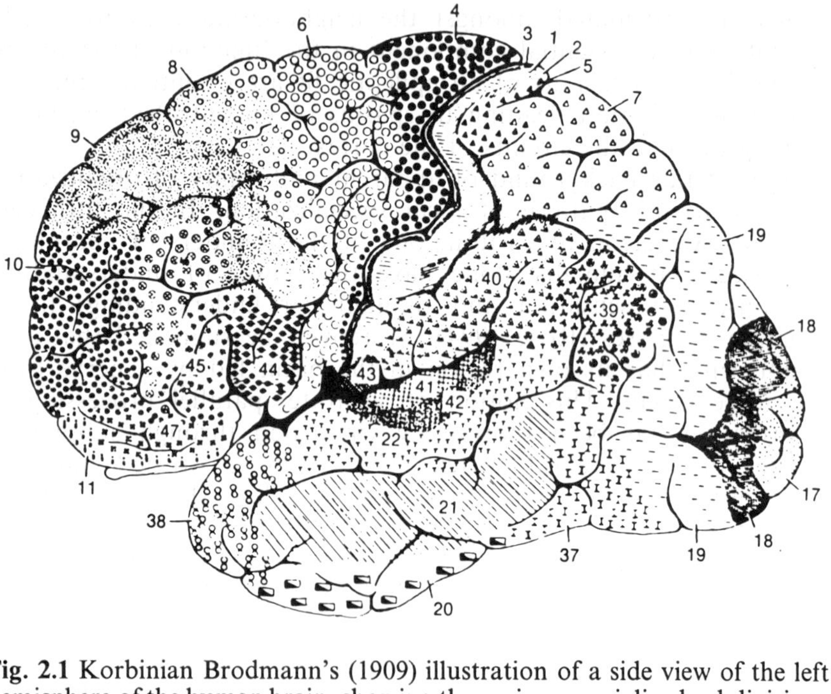

Fig. 2.1 Korbinian Brodmann's (1909) illustration of a side view of the left hemisphere of the human brain, showing the various specialized subdivisions that he recognized in microscopic sections of the cerebral cortex, on the basis of differences in the thickness of layers and the numbers of different cell types.

information processing (see Felleman and Van Essen 1991; Young *et al.* 1994).

Many of the functional subdivisions of the cortex coincide with subtle local variations on the common theme of cortical layering. The great German anatomist Korbinian Brodmann (1909), looking in his microscope at simply stained sections of the cerebral cortex, was able to delineate distinct regions, differing in the relative thickness of cell and fibre layers, and the proportions of different cell types, with sharp boundaries between neighbouring regions (Fig. 2.1). It is a testimony to Brodmann's powers of observation that the numbers he used to specify each cortical area are still widely employed to describe many of the major divisions of the cortex, in a wide variety of mammals, from rodents to human beings.

The regional diversity in the microscopic appearance of the cortex reflects local differences of circuitry between cortical neurons as well as variation in the proportions of different classes of neurons, projecting to different regions of the brain.

Brodmann himself recognized the likely relationship between anatomy and physiology. He wrote: 'The specific histological differentiation of cortical areas proves irrefutably their specific functional differentiation' (Brodmann 1909). Obviously the function that a cortical area performs depends on the input that it receives. It is no surprise, then, that distinct cortical areas receive axons from particular regions of the thalamus, hence gaining their input from particular sense organs or motor nuclei. This input in turn determines the functional properties of the area.

The formation of the cortex

It is a simplification, but not a gross one, to see the development of the cortex as occurring in two phases. The first, occurring prenatally in most mammals, is necessarily governed by the genes, slavishly reading out their instructions to make a roughly-formed brain. The second stage is a period of refinement in organization that is partly controlled by activity, reaching the cortex through the input from the thalamus, and through the web of connections within the cortex and between cortical areas. Thus, in this second phase, which mainly occurs postnatally, input from the sense organs could potentially influence the course of development of the brain and therefore mould the computing properties of cortical circuits to match more effectively the characteristics of the environment.

Building a kind of brain

Between about 2 and 4 months after conception of a human embryo, there is a phase of frenetic activity in the fluid-filled tube that is going to become the brain. Lining the walls of the tube is a epithelial sheet of *progenitor cells*, the ancestors of the neurons and glial cells in the brain. These progenitors divide rapidly, at first creating more of themselves and hence causing the tube to grow larger within the developing head. Then, at a certain point, different for each region of the brain, some of the daughter cells resulting from division of progenitors become committed to particular fates, destined to be certain kinds of glia or neurons (see Uylings *et al.* 1990). For a nerve cell, this commitment involves sacrificing the ability to reproduce. Once a daughter cell treads the path towards neuronhood, it loses the capacity to divide. Except in a few exceptional parts of the nervous system, all creation of neurons ceases after the initial developmental phase, and none can be replaced later in life. Quite unlike any other organ in the body, the brain is incapable of repair by replacement.

In the bulging wall of the top end of the developing tube, which will give rise to the cerebral hemispheres, committed daughter cells are produced at prodigious rates of up to a quarter of a million per minute in the human fetus. They crawl up, along the spindly bodies of specialized radial glial cells that extend from the inner to the outer surface of the tube (see Rakic 1981). The newborn neurons gather at the outer edge of the wall of the tube, gradually forming the layers of the developing cortex (Fig. 2.2).

At this stage of furious activity, the brain is particularly vulnerable to extreme influences from the outside environment. For instance, studies of children who were *in utero* at the time that atomic bombs were dropped on Hiroshima and Nagasaki, have shown that those who suffered serious neurological sequelae were all at a gestational age of 2–4 months at the time of their exposure to radiation. Equally, high levels of alcohol in the maternal bloodstream during that 2–4 month critical period can lead to the condition called Fetal Alcohol Syndrome in which the baby's cerebral hemispheres are small and the cortex is covered with a layer of tangled cells and fibres, presumably due to a failure in the normal migration and settling of cells entering the cortex.

The process of migration and the formation of the cortex has been studied in minute detail in a variety of species, but I shall describe the process in the rat. Compared with human beings and even monkeys, rats have tiny brains. Even in an adult rat, the cerebral hemispheres are smaller than a walnut and nowhere near as wrinkled. While the human cortex may have well over a hundred distinct functional areas, that of the rat probably has only a dozen or less. Consequently rats do not understand language, do not plan for the future except in the most rudimentary manner, probably see the world in very simple ways and presumably have little in the way of thoughts and intentions. Yet there is no reason to believe that the mechanisms for building the cortex are fundamentally different in rats and people. They are a lot easier to study and to understand in the rat!

Generally speaking, the waves of migrating cells accumulate in

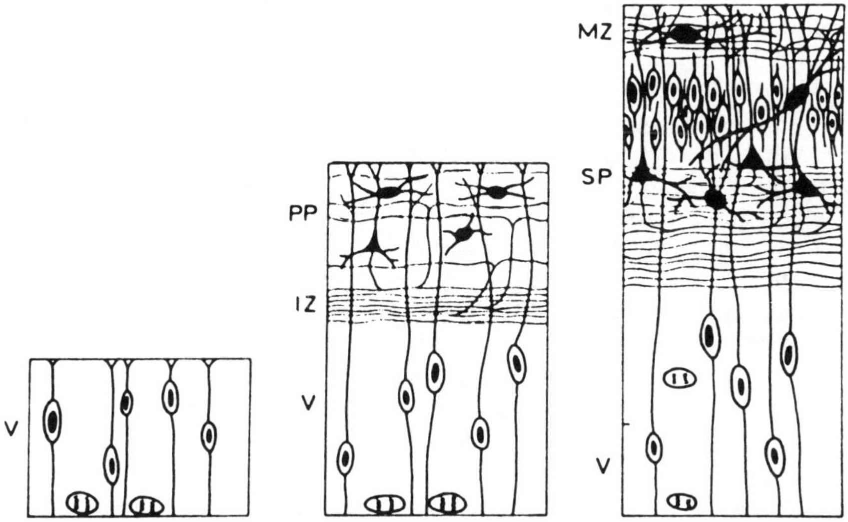

Fig. 2.2 Stages in the formation of the layers of the cerebral cortex. Each panel is a diagram of the whole thickness of the wall of the embryonic forebrain. At the bottom is the surface of the fluid-filled ventricle: at the top is the outer surface of the brain. Initially (left panel), before about the 12th day of gestation in the rat, there is only an epithelium of proliferating progenitor cells, forming the *ventricular zone* (V). By about the 15th day (middle panel), the first neurons have been produced and have migrated up to form the *preplate* (PP), with a layer of fibres forming the intermediate zone (IZ) below them. Somewhat later (right panel), still before birth in the rat, the nerve cells that will make up the main layers of the cortex have been born and are taking their place, in an inside-out sequence, between the outer part of the preplate (now called the *marginal zone*, MZ) and the lower part (called the *subplate*, SP). Nerve fibres from the thalamus have already arrived at this stage, but they are mainly accumulating in the subplate layer below the cortex proper, waiting for the cortex to become growth-permissive. (Modified from Uylings *et al.* 1990).

an inside-out sequence, the first forming the lower layers of the adult cortex. There is, though, a curious exception to this general rule. The very first neurons, born between about 11 and 14 days after conception in the rat (gestation lasts 21–22 days), amass to form a loose array of spindle-shaped cells called the *preplate*. Then, from about the 13th day of gestation, the arriving neurons migrate through the bottom part of the preplate and stop below the outer part of the preplate, forming a kind of neuronal sandwich (see Fig. 2.2). Subsequent waves of neurons repeat this process, gathering in inside-out sequence to form the layers of the cortex. Thus, some cells of the original preplate are left at the surface of the cortex, in what will become layer 1. The rest, called the *subplate*, come to lie underneath the thickening layers of the cortex proper.

The unusual cells of the original preplate are distinctive in a number of ways, quite apart from the pattern of their distribution. First, they become mature very rapidly, forming fully developed dendrites and turning on genes to produce a variety of transmitter chemicals. They send out axons, not only into the cortical plate above, but downwards towards the developing thalamus below. Indeed those descending fibres are the first to venture into the space between cortex and thalamus: hence they are known as pioneers. But perhaps the most puzzling characteristic of those first-born preplate cells is that many of them, at least in higher mammals, die after the cortex is formed. Their precocious maturity and early death obviously imply that preplate cells play a special role during development.

One can ask how much of the fate of a neuron in the cortex is determined at an early stage, even before the layers of the cortex are completed and activity arrives from the incoming thalamic fibres. One approach to this question is to infect progenitor cells with a virus that has been genetically engineered to produce a chemical inside the infected cells, which can be revealed with an appropriate chemical procedure applied to sections of the brain (see Walsh and Reid 1995). If a tiny dose of virus is injected into the fluid inside the developing brain, a small number of progenitor cells become infected and they transfer the infection to all their subsequent progeny. Such experiments have shown that, at early stages, individual progenitors can give rise to both neurons and glia among their daughters, but that later on, they become progressively more specialized in the types of cells they produce. A single progenitor produces many generations of daughters and hence contributes cells to several different layers of the cortex. Therefore, since cells in different layers send their fibres to different parts of the brain, a single progenitor must be capable of producing neurons that are distinctively different in this respect.

One might imagine, then, that some of the characteristics of a neuron, in particular the place to which it sends its axon, might be triggered by some aspect of the environment in the cortex at which the cell finds itself as it completes its migration. Ingenious experiments by Susan McConnell and her colleagues (see McConnell 1989) suggest that more properties are specified at the moment of birth of each neuron than might be imagined. She transplanted small fragments of progenitor epithelium from a young (ferret) embryo into the epithelium of an older animal. The host animal was of such an age that the lower layers of the cortex had already formed and the epithelium was generating cells bound for the upper layers. On the other hand, the transplanted tissue came from an embryo that was in the process of making cells destined for the lower layers. Although daughter neurons from the transplanted tissue were migrating into a cortex in which the lower layers had already formed, the majority of cells derived from the transplant jumped off the radial glia as they passed through the lower layers and took their place there, in the layers appropriate to the age of the donor embryo.

This remarkable result implies that the daughter cells have some way of identifying the cortical layer corresponding to the age of their maternal progenitor cell regardless of the unusual environment in which they find themselves. On the other hand, the same maternal cell produces daughters for different layers at different times during development. Therefore, whatever the layer-specific 'label' carried by each daughter, the label must change from daughter to daughter during normal development.

Although McConnell's experiments point to remarkably precise specification of some aspects of the fate of neurons at the time of cell division, it is clear that the local environment in which an immature neuron finds itself can influence its fate enormously (that is, the genes that it expresses and hence the specialized characteristics that it adopts). For instance, Ron McKay and his colleagues have shown that if uncommitted progenitor cells from one part of the brain are injected into a different region of the brain, they will differentiate and take on the appearance and characteristics of the neurons among which they find themselves (Vicario-Abejón *et al.* 1995), even distributing their axons to targets appropriate for the neurons surrounding them.

How are cortical areas created?

Since the newborn neurons climb up along the radial glia to reach their destination in the cortex, one might imagine that the defining differences between different functional areas in the cortex are specified by the characteristics of the progenitor cells that lie in the epithelium below each area. Take, for instance, the fact that layer 4 is relatively thick in primary sensory receiving areas (visual cortex, auditory cortex, and so on) compared with areas of cortex that control movement. The progenitor epithelium lying underneath regions that are destined to become primary sensory areas might simply generate more of the small cells that constitute layer 4.

Pasko Rakic was the first to suggest that the germinal epithelium consists of a kind of 'proto-map' of the cortex that it is going to make, each region being specialized to produce the appropriate proportions of different types of neurons in the correct sequence, so as to generate a particular type of cortex (see Rakic 1981). While the existence of the radial glia, apparently tying the germinal layer to the cortex above, seems to support this hypothesis, it is rendered less probable by the recent discovery that many of the progenitors are itinerant: they wander through the epithelium, some of them journeying along the entire length of the hemisphere, discharging daughter neurons into many different regions of cortex (Walsh and Reid 1995). For this whole process to be orchestrated so as to produce all the distinctly different cortical regions, with such sharp boundaries between them, seems implausible.

In fact, the distinctive anatomical characteristics of cortical areas do not become apparent until late in the process of formation of the cortex. At early stages, the entire cortex has the appearance of a fairly uniform sheet of tissue, with little or no hint of the areas and boundaries that will eventually form. The blandness of the appearance of primitive cortex has led to an alternative view of the origin of cortical areas – the so-called 'protocortex' hypothesis (O'Leary 1989). This suggests that some local environmental signal, imposed on each region of cortex, triggers regional differentiation and sets the boundaries of each area.

If we ask what could constitute such a local signal, there is only one candidate that can be seriously considered. The signal must be very local and precisely distributed (therefore it cannot be some hormone or other substance in the blood stream). There must be as many variants of the local signal as there are cortical areas. And the signal must be delivered during development of the cortex, before the first signs of specialization begin to appear. The only thing that is distributed so specifically to each region of the cortex, and at an early enough stage, is the array of fibres from the thalamus that will eventually bring signals into the cortex from the sense organs and other parts of the brain.

In the adult animal, the thalamus is itself subdivided into large numbers of cell groups, called *nuclei*. Simply speaking, each nucleus, which receives input from one of the sense organs or from some particular structure elsewhere in the brain, sends its axons to one particular area in the cortex. Indeed, the sharp boundaries of an area, visible in terms of the anatomical appearance of the cortex, correspond quite precisely to the limits of distribution of the axons from the particular thalamic nucleus that innervates that area. Either the extent of the cortical area and the distribution of thalamic axons are being identically regulated by some other factor or factors; or they are causally related, one to the other. In fact, mounting evidence suggests that the distribution of thalamic axons from each particular thalamic nucleus sets the boundaries of its individual cortical area and imposes other local specializations on that area, which contribute to its distinctive appearance.

Perhaps the most dramatic line of evidence comes again from experiments involving transplantation of tissue from one region of cortex to another (Schlaggar and O'Leary 1991). In the normal rat cortex, the posterior tip of the hemisphere is the primary visual cortex, which receives its input from the nucleus in the thalamus – the *lateral geniculate nucleus* – to which the eyes send their fibres. On the other hand information coming up from the skin of the body is distributed, via a different set of thalamic nuclei, to a strip of cortex in the middle of the hemisphere – the primary somatic sensory area. The lowest part of this somatic sensory area has an especially distinctive microscopic appearance in rodents, which mirrors the input that it receives from the whiskers of the upper lip. In layer 4 of this part of the cortex, the small cortical neurons are arranged into patterns called *barrels* (because their three-dimensional shapes resemble barrels). The closely-packed cortical cells constitute the 'walls' of each barrel, and the interior of the barrel is filled with a large bundle of thalamic fibres, forming synaptic contacts on the dendrites of the cortical cells, which point into the centre of the barrel.

In microscopic sections cut through layer 4, parallel to the layer, each barrel has the appearance of a ring of cells, the whole barrel being about a third of a millimetre across (Fig. 2.3). There is a regular pattern of barrels, arranged in rows across the cortex; and the number and

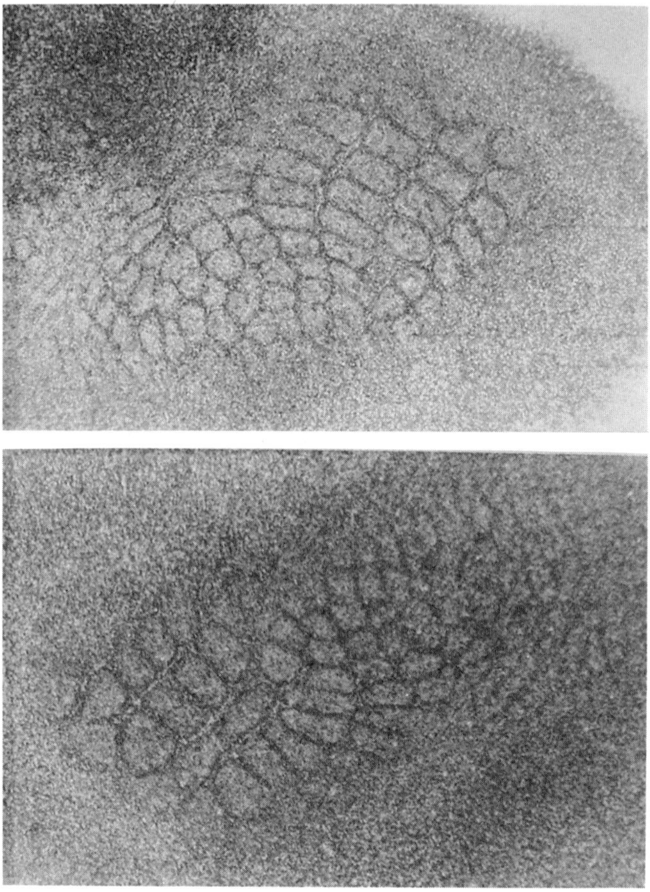

Fig. 2.3 These photomicrographs are of thin slices from the cerebral cortex of the mouse, cut parallel to the cortical surface, through layer 4 of the cortex, where the fibres from the thalamus arrive. This particular region of the cortex receives thalamic fibres carrying signals from the whiskers on the upper lip. Each tiny dot is the cell body of a cortical neuron in layer 4 and they can be seen to be arranged in rings or 'barrels', about 0.3 mm across. In the cell-free area in the middle of each barrel is a bundle of thalamic fibres, contacting the cortical cells in the wall of the barrel. The array of barrels corresponds precisely to the pattern of whiskers and it is clear that the cells of each particular barrel are receiving information mainly from one particular whisker. The upper photograph comes from a normal adult mouse, and the lower one from a mouse in which one row of whiskers was plucked on the day of birth (before the barrels have appeared). Barrels fail to form in the region of cortex corresponding to the plucked row of whiskers (lower left part of the bottom photograph), strongly implying that signals from the thalamic fibres are controlling the organization of the cortex early in life. (These photomicrographs were a gift from the late Hendrick Van der Loos).

distribution of these barrels exactly matches the number and pattern of whiskers on the lip (Welker and Van der Loos 1986). Since the barrels cannot really be discerned when axons from the appropriate region of the thalamus are growing into the somatic sensory cortex, around the time of birth, but they can be seen a few days later, it is parsimonious to conclude that the arrival of the thalamic axons triggers, in some way, the organization of the cortex into barrels. Powerful evidence for this hypothesis comes from the fact that in mice reared with one row of whiskers plucked on the day of birth, the field of cortical barrels develops abnormally, with the corresponding row of barrels failing to form (Van der Loos and Woolsey 1973: also see Fig. 2.3).

In certain mutant mice that have a tendency to form one or more extra whiskers, but not always the same number, and not always in the same position on the upper lip, an equivalent number of additional barrel structures always appears, at the matching position in the whole field of barrels (Welker and Van der Loos 1986). It is surely inconceivable that a single mutation could specify the development of both an extra whisker on the face and an extra grouping of cortical nerve cells to make a corresponding barrel. The conclusion must be that each whisker, acting through the nerves entering the cortex, induces the formation of a barrel.

O'Leary and his colleagues have also shown that other regions of cortex, though they do not normally do so, are capable of forming barrels if they are forced to accept axons from the thalamic nucleus that receives input from the whiskers (see O'Leary *et al.* 1995). In newborn rats, they transplanted pieces of cortex from the back of the hemisphere (which should therefore have become visual cortex) into the region that should have formed barrels. *Mirabile dictu*, the erstwhile visual cortex soon organized itself to take on the appearance of barrels, as the appropriate thalamic fibres grew into it. This remains the best evidence that the distinctive regional differences in cortical organization that Brodmann described are at least partly triggered by the particular type of thalamic axons that invade the developing cortex, and by the kind of signals that they are relaying from the sense organs.

How do thalamic axons know what to do?

If, as seems increasingly likely, the arrival of thalamic axons constitutes a signal that triggers at least part of the process of regional differentiation, then the mechanism by which those axons are guided to the appropriate part of the cortex is crucially important in determining the future of the cortex. This is, of course, just a special instance of the general question: how is the growth of axons controlled?

On the basis of studies of a variety of pathways in the brain and spinal cord, developmental neuroscientists have long recognized four mechanisms that might enable an array of growing axons to be guided to form a topographically ordered distribution over a target structure (Udin and Fawcett 1988).

1. The target structure might have an array of chemical 'addresses', perhaps in the form of a gradient of concentration of some substance produced by the target cells, which is recognized by the growing axons, constituting a kind of chemical lock-and-key mechanism.
2. Axons from a group of cells might grow out as an orderly array and simply maintain their spatial relationships so that they distribute themselves over the target in a pattern that reflects their points of origin.
3. In a variant of this notion of fibre ordering, the axons could grow out in a temporal sequence, reflecting a wave of maturation sweeping over the group of cells from which they originate, and take their place across the target structure in a first-come, first-served distribution. Since time is uni-dimensional, such *chronotopicity*, as it is called, could account only for one axis of a two-dimensional map of fibres on to their target.
4. Fibres might be initially distributed in a more-or-less disordered pattern and sort themselves out by local growth of axon terminals in the target structures, using information derived from chemical or electrical signals that they carry to determine where they end up.

Recent work has emphasized another general mechanism of axon guidance, namely interaction of the growing fibres with markers, signposts, and other guidance signals along the way (see Goodman and Shatz 1993). Substances deposited by cells along the pathway into the extracellular space can attract or repel growing axons, hence steering them along a particular course. In addition, clusters of cells strategically placed along the way can act as beacons, helping to guide the navigating fibres. And finally, axons can grow selectively along the surfaces of pre-existing axon bundles, which have 'pioneered' pathways within the developing brain. The use of pioneer axons as guidance scaffolds, which is well established in invertebrates, is an attractive notion, but it raises the question, of course, of how the pioneers are themselves guided.

Although the field is not without controversy, a picture is emerging of the process by which thalamic axons from each nucleus reach, with considerable precision, the appropriate region of the cortex. Two quite new techniques have energized this area of study: tissue culture, to maintain tiny living fragments of brain tissue in 'test-tube' conditions; and fluorescent dyes that can stain growing nerve fibres, either in such cultured brain fragments or in whole embryonic brains that have been chemically fixed for microscopic examination.

Zoltán Molnár and I have used the culturing method to try to produce a simplified 'model' system in which to study the mechanisms involved in axon guidance (Molnár and Blakemore 1991, 1995). We placed side-by-side in a culture chamber a slice of cerebral cortex, taken from an anaesthetized rat fetus or very young postnatal animal, and a tiny explant dissected from the developing thalamus of a fetus on the 16th day of gestation (the age at which axons are normally growing out towards the cortex). We soaked the thalamic explant in a solution of a fluorescent dye (called 'DiI'), so that any axons sprouting from the explant could be followed in a fluorescence microscope. In our initial experiments, we always took the thalamic explant from the region of the lateral geniculate nucleus (the nucleus that receives input from the eyes) and the cortical slice from the posterior part of the hemisphere, the putative visual cortex, to which the lateral geniculate nucleus sends its fibres in the living fetus.

Whatever the age of the animal from which the slice of cortex is taken, axons do indeed grow out from the fragment of thalamus; and there is much less profuse outgrowth in the absence of a piece of cortex. So, the first conclusion is that the cells of the cortex, even at the earliest stage, produce a soluble *growth-promoting* substance, which diffuses through the culture medium and stimulates the outgrowth of axons from the thalamus.

If the slice of cortex comes from a rat fetus younger than the 19th day of gestation, the thalamic fibres do not actually penetrate the slice of cortex but grow around it and over it: at this stage the tissue of the cortex itself will not allow the ingrowth of thalamic axons. On about the 19th day (2–3 days before birth) a gene appears to be suddenly switched on in the cortex, producing *growth-permissive* molecules on the membranes of cortical cells, making their surfaces attractive to the ingrowth of thalamic fibres. Fibres from the thalamic explant grow into a slice of cortex taken on the 20th day; but, instead of stopping and terminating on cells in layer 4, they grow all the way through the cortex and then ramify over its surface. However, if the cortical slice comes from a rat pup, 2 days or more after birth, the thalamic fibres behave remarkably normally, growing up through the lower layers of the cortex, but slowing down, branching and forming synapses on cells in what is presumably destined to be layer 4 of the cortex. Thus, there appears to be another molecular signal (that is, another pattern of gene expression) that turns on about 2 days after birth among the cells of layer 4, which somehow encourages invading thalamic fibres to terminate. We have called the latter a *stop signal* (Molnár and Blakemore 1991).

Thus the cortex produces a sequence of molecular signals that could influence, even determine, the timing of outgrowth, invasion, and termination of thalamic fibres (Molnár and Blakemore 1995). We have studied the chronology of these events in the intact, developing brain by placing tiny crystals of fluorescent dye into parts of the whole brain, of various ages, fixed chemically to preserve the tissue, which was then cut in sections to view in a fluorescence microscope. The dye diffuses along nerve fibres, revealing them like thin rootlets, growing through the developing brain, searching for their targets in other parts of the brain.

When such a crystal of dye is implanted directly into the thalamus of a rat fetus around the 16th day of gestation, it stains axons that have already grown, in a wonderfully ordered pattern, towards their correct regions of the cortex. Fibres from the visual nucleus in the thalamus, first grow down to the base of the thalamus and then sweep out sideways through the *internal capsule* – the narrow portal that connects the thalamic region to the cerebral hemispheres. They then curve upwards and run unerringly towards the back of the hemisphere, the region that is destined to become the primary visual part of the cortex. They start to grow on the 14th day of gestation, just as the cortex itself is beginning to form, and, according to the tissue culture experiments, is already generating the diffusible substance, capable of stimulating the sprouting of axons from explants of the thalamus *in vitro*. It seems likely that this growth-promoting factor plays a part in initiating the outgrowth of thalamic fibres *in vivo*.

It takes only two days for those growing thalamic fibres to reach the visual region of the cortex; but at that age, 16 days after conception, the tissue culture experiments suggest that cells of the cortex do not have the right surface properties to support the ingrowth of axons. In the living embryo the newly arrived thalamic fibres do not substantially invade their region of cortex. They gather, in the order in which they arrived, like a British bus queue, waiting for the cortex above to let them come on board. At least some of the thalamic fibres actually form functional synaptic contacts on those mysterious early-born cells of the subplate, among which they accumulate (see Goodman and Shatz 1993). Who knows what dialogue may be going on between those thalamic axons and their curious targets, or what role this interaction may have in the ultimate sculpting of the cortex?

The process by which the ordered array of fibres from the entire thalamus distributes itself in the correct pattern over the whole sheet of the cortex cannot be dependent on signals from the sense organs, since the axons from the eyes, ears, and so on, have not themselves navigated towards their appropriate thalamic nucleus at the time that thalamic fibres set off on their expedition towards the unknown of the cortex. The basic form of the network of input to the cortex is apparently determined by an autonomous interaction between cortex and thalamus.

Quite suddenly, around the 20th day of gestation in the rat, a couple of days before birth (just when cortical slices *in vitro* become permissive to ingrowth), the waiting thalamic fibres detach themselves from the subplate layer, and grow up, at about 0.5 mm per day, into the grey matter of the cortex above. A few of them manage to pass through the whole thickness of the cortex and establish connections with uppermost layer of the cortex. But the vast majority arrive amongst the immature cells destined to become layer 4 of the cortex a couple of days after birth, just as those cells begin to express the stop signal revealed by the tissue culture experiments.

Thus the crucially timed steps in the navigation of those fibres that are going to provide the input of information to the cerebral cortex could be programmed by a sequence of simple molecular signals generated by the switching on of genes in the cortex itself. The diffusible growth-promoting substance might help to stimulate the initial outgrowth and influence the direction of growth; onset of the growth-permissive property could play a part in determining the end of the 'waiting period' and encouraging the invasion of the cortex itself; and the stop signal of layer 4 is produced right on cue, persuading most of the thalamic fibres to form their connections on exactly the correct target cells.

How do thalamic fibres know where to go?

Cast your mind back to those various mechanisms that have been proposed whereby fibres might be guided to create a topographic distribution across a target structure (Udin and Fawcett 1988: see above), of which the projection of the thalamus on to the sheet of the cerebral cortex is the most spectacular example. The clear evidence from tissue culture experiments that the cortex produces a sequence of chemical signals capable of influencing thalamic fibres obviously raises the possibility that one or more of those substances act as markers that define the appropriate region of cortex for each group of thalamic fibres.

In the living embryo, fibres from the lateral geniculate nucleus, the visual part of the thalamus, make their way almost faultlessly to the occipital part of the cortex and none of them strays into the alien reaches of the frontal cortex, for example. Could it be that the growth-promoting substance, diffusing down from the cortex, could vary in its chemical nature from one region of cortex to another, each attracting only axons from the correct region of the thalamus? This seems utterly implausible: the mixing of chemicals from different cortical areas, as they diffuse towards the thalamus is incompatible with the distinct ordering of axons as they descend from the thalamus, downwards and sideways towards the internal capsule. Nevertheless, Molnár and I tested this theory experimentally by culturing together a single fragment of thalamus with two entirely different regions of cortex, one the correct target, the other a region never normally innervated by fibres from that part of the thalamus. The results were very clear: the growth of thalamic fibres towards and (if the cortex came from a fetus of an age at which the growth-permissive properties of the cortex had turned on) into the cortex was indistinguishable in the two cortical slices (Molnár and Blakemore 1991).

These *in vitro* experiments argue that the growth-promoting and growth-permissive properties do not provide chemical clues about *position* within the target. We must look to one of the other possible mechanisms of guidance or, more likely, to a combination of several of them. Certainly there does seem to be a temporal gradient of maturation across the developing thalamus and it is likely that fibres grow out in sequence, depending on the position of their cell bodies within the thalamus. Thus chronotopy may help to determine one axis of the mapping on to the cortex. Use of the dye staining technique to observe thalamic fibres in the embryonic brain shows that they maintain remarkable ordering as they grow towards the cortex. But most interesting of all is the evidence that pioneering fibres might help to coax the thalamic axons towards the correct region.

It turns out that those precocious, early-born cells of the subplate, which lie below the thickening cortex, send out axons of their own, downwards from the cortex towards the internal capsule, at the same time that fibres are sprouting from the thalamus (Fig. 2.4). These descending fibres pioneer the route within the cerebral hemispheres and they meet the array of thalamic fibres as they emerge through the internal capsule. Molnár and I have shown that axons from corresponding regions of cortex and thalamus meet, in what we have called a 'handshake', and that the thalamic axons grow over the pioneering subplate axons as they make their way up to wait below the correct region of cortex (see Molnár and Blakemore 1995). Clearly this intimate association between descending and ascending fibres could play a part in the guidance of the latter.

THE SECOND STAGE IN THE BUILDING OF THE BRAIN: THE IMPRESSION OF PERSONAL EXPERIENCE

We see that the initial construction of the cerebral cortex, with its layers and basic cell types, is accomplished at an early stage, before birth in placental mammals. The elegance of the mechanism that controls this initial phase, depending on a cascade of individually simple mechanisms and interaction of growing fibres with the internal environment in which they find themselves, is impressive. One begins to see how the very limited information available in the genes could orchestrate the construction of a basic kind of brain.

But once the sense organs have connected to their own thalamic areas, once the thalamic fibres have invaded the cortex, new mechanisms of regulation are possible. Activity, in the form of nerve impulses propagating along the pathways of fibres, might influence the course of development. Initially such activity may have nothing to do with the external environment. The eye obviously cannot function until the photoreceptors are able to catch light in the retinal image and convert it into signals that are transmitted to the brain; these photoreceptors do not become functional until some time after birth in many mammals – usually a short time before the eyes open. In cats, for instance, the photoreceptors do not start working until a few days before the eyelids open, about 7–10 days after birth. However, there is activity in the retina before that, helping the cortex to sort out and refine the crude connections that have already been made. Carla Shatz and her colleagues have shown that waves of activity sweep slowly across the retina in young animals, before the eyes have opened (see Goodman and Shatz 1993); so patterns of impulses are transmitted to the brain, and the timing of those impulses is an

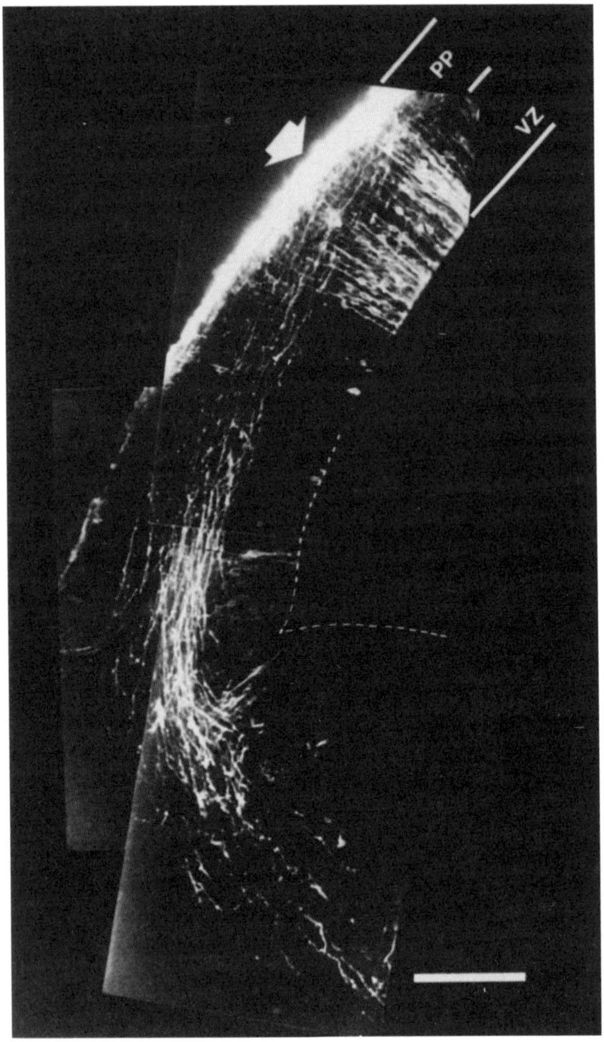

Fig. 2.4 The brain of a fetal mouse (about 13 days after conception, equivalent to the 14th day in a rat fetus) was chemically fixed to preserve it and a tiny crystal of a fluorescent dye was inserted into the surface of the cortex. This figure is a photomicrograph of a cross-section through the wall of the developing cerebral hemisphere. At this stage, the germinal cells of the ventricular zone (VZ) have given birth to their first generations of neurons, which have migrated up along the processes of radial glial cells to form the preplate (PP) (see Fig. 2.2). The bright dye has labelled these cells directly below the point at which the crystal was placed, the position of which is

indicator of the proximity of the cells of the retina that give rise to them. The thalamus, and perhaps the cortex too, probably make use of the information about proximity to refine the distribution of fibres from eye to thalamus and from thalamus to visual cortex, even before the animal has seen anything of the world around it.

Sensitive periods: learning from the world

Consider again the example of the barrel structures in the body representation in the cortex, which are not visible around the time of birth but which appear after thalamic fibres grow in (and can be induced in inappropriate parts of cortex transplanted from other regions). The formation of these barrels during the first few days after birth, in rats and mice, is a compelling example of a general phenomenon: regions of cortex, especially the main sensory areas, pass through distinct periods, during which particular characteristics develop, and these phases of development usually require input from the sense organs to maintain normal development.

Take the case of the primary visual cortex, the much studied region at the very back of the cerebral hemispheres. In adult cats and monkeys, most individual nerve cells in this area have two physiological properties quite different from those of the thalamic fibres that provide their input. Whereas the thalamic fibres respond with a burst of impulses when a particular point on the retina is stimulated with a bright or dark spot, cortical neurons in the primary visual cortex respond strongly only if a *line* or *edge* of a particular orientation appears at the right position in the visual field (see Hubel 1988). Moreover, they usually respond if either eye is stimulated, with the same orientation, and even more strongly if both eyes are simultaneously stimulated, as they would be when viewing a normal object at a particular distance from the eyes (Barlow *et al.* 1967). The properties of orientation selectivity and binocularity, which presumably play a part in the analysis of the shapes and distances

---

indicated by a white arrow. In addition the fluorescent dye has stained a bundle of 'pioneering' nerve fibres growing out of the preplate cells, even at this early stage, and down towards the bottom part of the cerebral hemisphere. These descending fibres meet the corresponding axons from the thalamus growing upwards towards the cortex. Zoltán Molnár and I have suggested that the thalamic fibres complete their growth over the array of descending fibres, which might therefore help to guide them towards the correct region of cortex. The horizontal scale bar at the bottom is 0.1 mm. (Photomicrograph prepared by Zoltán Molnár and Richard Adams.)

of objects, are created at the level of the cortex: they must depend, in some still uncertain way, on the precise pattern of connections from the incoming thalamic fibres and their cortical target cells, as well as on the rich interconnections within the cortex itself.

The property of orientation selectivity is itself mapped across the cortex, on a microscopic scale. If a microelectrode is introduced into the visual cortex of an anaesthetized cat or monkey, and driven down perpendicular to the surface, all the neurons recorded along the way, through the full 2 mm thickness of all the cortical layers, respond best to lines of virtually the same orientation. If the electrode is taken out and inserted again, for example at a distance of 0.05 mm, all the cells encountered during the penetration will prefer a slightly different orientation. Thus, the neurons of the cortex form *columns* with shared stimulus preference and, across the surface of the cortex, there is a usually continuous, progressive shift in preferred orientation from column to column (see Hubel 1988). Each patch of cortex, less than 1 mm across, contains the neural 'computing' machinery to detect any orientation of contour falling at a particular position on the retina.

Given the remarkable precision of this organization, it is surprising to discover how much of it is pre-programmed before the onset of visual experience. In the visual cortex of a week-old kitten, just before the eyes open, or in a newborn monkey, many neurons of the primary visual cortex already respond selectively (but less precisely than in the adult) to lines or edges; many of them already respond through either eye. But it is clear that both of these properties are maintained and refined as a result of the animal's experience during the first few weeks of vision.

Over the first few days after a kitten opens its eyes, there is a rapid increase in the fraction of orientation selective neurons, and the whole population becomes more precisely 'tuned' and more vigorously responsive (for example Blakemore and Van Sluyters 1975). On the other hand, if a kitten is reared for a few weeks in the dark, the proportion of cortical cells responding selectively to lines and edges decreases progressively and the cells become generally rather sluggish or even entirely unresponsive.

Vision during a short time early in life is, then, needed to sustain the normal maturation of the visual cortex. However, more than just validating a predetermined (presumably genetically determined) program of development, sensory experience can actually steer the direction of neuronal specialization in an adaptive way.

If a kitten is deprived of vision in one eye during the first few weeks of life (by closing the eyelids or putting a translucent contact lens on the eye), even for as little as a few hours at 4 weeks of age, the majority of cortical cells lose their functional input from that deprived eye and respond only to visual stimuli shown to the other eye. One might conclude that the thalamic fibres bringing information into the cortex from that eye degenerate as a result of deprivation, but that is not the case. These relatively inactive fibres are still present, but they form smaller branching distributions among the cells of layer 4 than do those carrying signals from the other eye. Similar changes in the eye dominance of neurons can occur in the monkey's primary visual cortex, and in both species there is a fairly short sensitive period for these changes in functional connectivity (see Hubel 1988). Perhaps the best evidence that such shifts in the eye dominance of nerve cells is the fact that they can very rapidly be reversed if the deprived eye is reopened and the other one closed, as long as the animal is still young enough to be within its sensitive period (Blakemore and Van Sluyters 1974).

Almost certainly, similar effects can occur in the human cortex, since babies who have suffered a cataract, or just a focusing error in one eye, usually have very poor vision through that eye later in life (a condition known as *amblyopia*), even if the cataract or defocus is completely corrected. It is likely that the sensitive period for human vision starts at about 3–6 months and lasts until 1–2 years of age. Treatment of amblyopia by removing the cataract or correcting the focus of the affected eye and patching the other one can be remarkably effective if carried out early enough, just as in animals.

The actual angle of edges preferred by individual visual cortical cells can also be altered by selective visual experience and deprivation. If a kitten is reared in an environment that virtually prevents it from seeing horizontal lines, few cells in its cortex will subsequently respond to horizontal contours and the columnar structure becomes biased towards those angles of lines and edges that it saw when it was young (see Blakemore 1974).

How does the environment change the brain?

On reflection it should be no surprise to discover that the physiological properties of sensory areas of the brain can be modified by personal experience. After all, our brains are capable of being changed every second of every day of our lives, whenever we form new memories. If you look at a stranger's face and commit it to memory, you have a good chance of recognizing that person days, months, or even years later. Something must have changed inside your brain to enable you to know that a particular complex visual stimulus (the view of the face) is the same as one you have seen before, and, moreover, to drag up from the cellar of remembrance all the other recollections associated with that person's face – the context in which you first met them, the things you talked about, perhaps a little of their life story gleaned from that first conversation, and so on.

Although the way in which the actual content of remembered experiences is represented in the brain remains a puzzle, much has been learned of the probable cellular mechanisms that underlie not only the conscious memories of adulthood but also the reorganizations in sensory areas that occur early in life. Most excitatory nerve fibres, almost certainly including the thalamic axons that carry sensory signals into the cerebral cortex, secrete the transmitter substance glutamic acid (or glutamate) to communicate with the cells they end on. Transmitter substances act by diffusing across the minute gap between the axon terminal and the membrane of the next nerve cell, and interacting with complex protein molecules called receptors that are embedded in that membrane.

There are several different types of receptor that specifically detect and respond to glutamate. The main type of glutamate receptor is coupled to a pore in the membrane that, when activated, allows sodium ions from the external fluid to flow in, through the membrane, making the interior of the cell more positively charged, which tends to set off a nerve impulse in that receiving cell. However, another type of glutamate receptor, which glories in the name *N*-methyl-D-aspartate receptor (*NMDA* receptor for short), behaves rather differently. NMDA receptors are activated by glutamate only if the interior of the receiving nerve cell is already positively polarized through the activation of nearby conventional glutamate receptors. When stimulated, the NMDA receptor allows calcium ions into the cell, and these in turn set off a series of intracellular reactions that tend to strengthen the efficiency of that synapse.

The requirement that the receiving nerve cell should already be depolarized (through the activity of other synaptic connections on to that cell) means that synapses that include the NMDA receptor system can act as detectors of *coincidences* or correlations in firing of different incoming nerve fibres. The fact that activation of the NMDA receptor then leads to strengthening of that synapse means that the NMDA receptor system is capable of *learning* those coincidences or associations.

It turns out that the genes that produce NMDA receptors are not equally switched on in all nerve cells. They are especially active in certain parts of a brain structure called the *hippocampus*, tucked under the temporal lobe of the cerebral hemispheres, which has long been thought to be involved in the laying down of personal memories. It turns out that nerve cells in this structure are capable of undergoing a change called long-term potentiation: if the nerve fibres providing input to the hippocampus from the cerebral cortex are given a brief burst of electrical stimulation (a condition which would lead to the activation of NMDA receptors), the cells of the hippocampus become much more sensitive and responsive to subsequent stimulation of those nerves. This potentiation lasts hours, days, or perhaps for ever. The involvement of the NMDA receptor in the process of potentiation is strongly implicated by the fact that drugs that block the receptor prevent potentiation, and, in the living animal, they interfere with certain kinds of learning.

The NMDA receptor is found on the membranes of other nerve cells throughout the brain at certain times in life. In the visual cortex, neurons have a particularly dense coat of NMDA receptors in their synapses during the very sensitive period when those cells are known to be modifiable as a result of visual stimulation. Drugs that antagonize the NMDA receptor will also prevent the modification of neuronal properties in the visual cortex produced by rearing an animal with, say, one eye covered up (see Fox and Daw 1993). This all suggests that certain forms of memory in the adult animal and modification of the sensory areas of the cortex during development probably depend on rather similar cellular mechanisms.

The adaptive value of having modifiable circuitry in the brain, influenced by the animal's environment, is obvious. If appropriate learning rules are implemented in the characteristics and positioning of the 'plastic' synapses, the changes in the properties of individual nerve cells and the circuits that they form could result in the gaining of information by that circuitry. For instance, the increase in the selectivity of visual cortical neurons for the angles of lines and edges, as a result of visual experience, should lead to more accurate encoding or representation of the shapes of objects seen by the eyes. Equally, the ability of cortical neurons to shift their actual preferences for orientation towards those angles of edges or lines seen most during early life could help to match the characteristics of the visual system to the statistical distribution of the angles of edges in the animal's particular visual environment.

The adaptive value of the rapid modifiability in eye dominance of visual cortical neurons is surely not to guarantee that one eye becomes virtually blind if vision is compromised through that eye early in life. It presumably has a more subtle function, enabling cortical neurons to change their exact preferences through the two eyes, on the basis of the habitual pattern of binocular stimulation during normal viewing, so as to fit the system of binocular fusion to the precise characteristics of the world.

It seems as if synaptic plasticity enables the brain to take advantage of information that can be provided by the pattern of sensory stimulation derived from the animal's environment, 'fine-tuning' the circuitry that has been roughly laid down on the basis of that cascade of simple genetic instructions occurring before birth.

In the same way, individual memories, skills, and other forms of learning in the adult animal provide the brain with representations of individual interactions with the external environment, expanding the information embedded in the brain, far beyond the knowledge of the world captured in the simple reflexes and reactions established by the genetically programmed initial wiring.

The appearance, during evolution, of mechanisms by which the strengths of synaptic connections could be changed as a result of stimulation, so as to capture information about the outside world and about individual experiences, was surely one of the most important and portentous steps in organic evolution. It seems to have occurred quite early in the history of life, since many invertebrates have the capacity to learn, and the mechanisms underlying their simple forms of learning are quite similar to those that occur in the mammalian hippocampus and cerebral cortex (see Kandel 1991).

From the Empiricist philosophers onwards, some have seen the role of experience as being distinctly separate from, even opposed to, innate influences on the nature of brain and mind – the classical issue of Nature versus Nurture. But consider where the fundamental basis of synaptic plasticity resides: not in some arbitrary and accidental effect of experience on the brain, but in a precisely orchestrated and naturally selected set of characteristics of nerve cells and their connections. The NMDA receptor is a protein, directly produced by genes, which have been selected, presumably because of the advantages bestowed on an animal that has modifiable synapses.

The actual changes that underlie the strengthening of a synapse probably involve a variety of mechanisms, including an increase in the amount of transmitter substance released by the incoming axon terminal, the production of new receptor molecules in the membrane of the receiving nerve cell, the actual growth and branching of the nerve fibre and expansion of its terminals, as well as the proliferation of the dendrites of the receiving cell and the formation of the specialized spine structures on which many synapses are formed. All of these changes would seem to require, ultimately, the mobilization of the genes, to create new receptor and transmitter molecules and to generate all the components of new dendrites, axon branches, and synaptic terminals. In short, the environment usually exerts its influences on the brain by modulating the activity of the genes. The genes have discovered a trick for harnessing the environment to control their own expression. Not Nature *versus* Nurture, but Nurture in the service of Nature!

REFERENCES

Barlow, H.B., Blakemore, C., and Pettigrew, J.D. (1967). The neural mechanism of binocular depth discrimination. *J. Physiol.*, **193**, 327–42.

Blakemore, C. (1974). Developmental factors in the formation of feature extracting neurons. In *The neurosciences: third study program* (ed. F.O. Schmitt and F.G. Worden), pp. 105–13. M.I.T. Press, Cambridge, Mass.

Blakemore, C. (1977). *Mechanics of the mind.* Cambridge University Press.

Blakemore, C. and Van Sluyters, R.C. (1974). Reversal of the physiological effects of monocular deprivation in the kitten: further evidence for a sensitive period. *J. Physiol.*, **237**, 195–216.

Blakemore, C. and Van Sluyters, R.C. (1975). Innate and environmental factors in the development of the kitten's visual cortex. *J. Physiol.*, **248**, 663–716.

Brodmann, K. (1909). *Vergleichende Localisationslehre der Grosshirnrinde in ihren Prinzipien dargestellt auf Grund des Zellenbaues.* J.A. Barth, Leipzig.

Clarke, E. and O'Malley, C.D. (1968). *The human brain and spinal cord: a historical study illustrated by writings from antiquity to the twentieth century.* University of California Press, Berkeley.

Felleman, D.J. and Van Essen, D.C. (1991). Distributed hierarchical processing in the primate visual cortex. *Cerebral Cortex*, **1**, 1–47.

Fox, K. and Daw, N.W. (1993). Do NMDA receptors have a critical function in visual cortical plasticity? *Trends Neurosci.*, **16**, 116–22.

Goodman, C.S. and Shatz, C.J. (1993). Developmental mechanisms that generate precise patterns of neuronal connectivity. *Neuron*, **10**, 77–98.

Hubel, D.H. (1988). *Eye, brain, and vision.* W.H. Freeman, Scientific American Library, New York.

Kandel, E.R. (1991). Cellular mechanisms of learning and the biological basis of individuality. In *Principles of neural science* (3rd edn) (ed. E.R. Kandel, J.H. Schwartz, and T.M. Jessell), pp. 1009–31. Appleton and Lange, East Norwalk, CO.

McConnell, S.K. (1989). The determination of neuronal fate in the cerebral cortex. *Trends Neurosci.*, **12**, 342–9.

Molnár, Z. and Blakemore, C. (1991). Lack of regional specificity for connections formed between thalamus and cortex in coculture. *Nature*, **351**, 475–7.

Molnár. Z. and Blakemore, C. (1995). How do thalamic axons finds their way to the cortex? *Trends Neurosci.*, **18**, 389–97.

O'Leary, D.D.M. (1989). Do cortical areas emerge from a protocortex? *Trends Neurosci.*, **12**, 400–6.

O'Leary, D.D.M., Borngasser, D.J., Fox, K., and Schlaggar, B.L. (1995). Plasticity in the development of neocortical areas. In *Development of the cerebral cortex.* (*Ciba Foundation Symposium* **193**) (ed. G.R. Bock and G. Cardew,), pp. 214–30. John Wiley and Sons, Chichester.

Rakic, P. (1981). Developmental events leading to laminar and areal organization of the neocortex. In *The organization of the cerebral cortex* (ed. F.O. Schmitt), pp. 7–28. MIT Press, Cambridge, MA.

Rose, F.C. (1994). The neurology of Ancient Greece – an overview. *J. Hist. Neurosci.*, **3**, 237–60.

Schlaggar, B.L. and O'Leary, D.D.M. (1991). Potential of visual cortex to develop arrays of functional units unique to somatosensory cortex. *Science*, **252**, 1556–60.

Udin, S.B. and Fawcett, J.W. (1998). Formation of topographic maps. *Ann. Rev. Neurosci.*, **11**, 289–327.

Uylings, H.B.M., Van Eden, C.G., Parnavelas, J.G., and Kalsbeek, A. (1990). The prenatal and postnatal development of rat cerebral cortex. In *The cerebral cortex of the rat* (ed. B. Kolb and R.C. Tees), pp. 35–76. MIT Press, Cambridge, MA.

Van der Loos, H. and Woolsey, T.A. (1973). Somatosensory cortex: structural alterations following early injury to sense organs. *Science*, **179**, 395–6.

Vicario-Abejón, C., Cunningham, M.G., and McKay, R.D.G. (1995). Cerebellar precursors transplanted to the neonatal dentate gyrus express features characteristic of hippocampal neurons. *J. Neurosci.*, **15**, 6351–63.

Walsh, C. and Reid, C. (1995). Cell lineage and patterns of migration in the developing cortex. In *Development of the cerebral cortex* (*Ciba Foundation Symposium* **193**), pp. 21–40. John Wiley and Sons, Chichester.

Welker, E. and Van der Loos, H. (1986). Quantitative correlation between barrel-field size and the sensory innervation of the whiskerpad: a comparative study in six strains of mice bred for different patterns of mystical vibrissae. *J. Neurosci.*, **6**, 3355–73.

Young, M.P., Scannell, J.W., Burns, G.A.P.C., and Blakemore, C. (1994). Analysis of connectivity: neural systems in the cerebral cortex. *Rev. Neurosci.*, **5**, 227–49.

3

Emotion and reason in the future of human life

Antonio R. Damasio

Antonio Damasio is Van Allen Distinguished Professor and Head of the Department of Neurology at the University of Iowa; and Adjunct Professor at the Salk Institute in La Jolla, California. He received both his MD and his doctorate in his native Portugal, at the University of Lisbon. His research team in Iowa has made outstanding contributions to the understanding of memory, language, emotion, and decision-making; with his wife, Dr Hanna Damasio, he has created at the University of Iowa one of the world's leading centres for the investigation of the neural basis of cognition. Professor Damasio is a member of the National Academy of Sciences' Institute of Medicine, the American Academy of Arts and Sciences, the National Advisory Council on Neurological Diseases and Stroke, and the European Academy of Sciences and Arts. His most recent book, Descartes' Error: Emotion and Reason in the Human Brain, *has been published worldwide.*

The title of this chapter may need some explanation, since the juxtaposition of the terms emotion and reason is unusual and the reference to the future of human life may be puzzling. First, emotion and reason have traditionally been considered polar opposites but modern neurobiological thinking suggests that they are not. In this chapter I present some of the findings that led to this idea, introduce a hypothesis concerning the role of emotion in the process of reasoning and decision-making, and discuss some preliminary testing of the hypothesis. Second, in this context, the reference to the future of human life should now be a bit more transparent. It is difficult to conceive of any future for human life without an abundance of collective human wisdom and such wisdom depends upon a well tempered machinery for decision-making within which emotion and reason are interwoven. Third, all the reader will have left to wonder about is the connection of these topics to the central theme of the Linacre Lecture series and of this book, that is, the environment. And here the answer is no less clear. The future of human life is inseparable from a concern for the environment, in the broadest sense of the term. The collective wisdom required for the continuation of human life is the same as that required for maintaining a survivable physical and social environment.

A CHALLENGING OBSERVATION

My interest in the neurobiological mechanisms of reason began with a series of observations in neurological patients, for which there was no adequate explanation and which thus posed a challenge (Damasio 1979). Until a specific form of brain damage occurs, these patients could be described as intelligent and successful. After the onset of their neurological disease and of their brain lesions (a lesion is an area of circumscribed damage), a pattern of abnormal decision-making in personal and social matters develops and the creativity and good adaptation that characterized the patients' lives until then is suddenly gone and gone for good. The patients have difficulty planning their future, be it the long range future or the immediate future of the following work day. No less importantly they also have difficulty choosing suitable partners in their social environment. The plans they organize, the people they elect to join with, or the activities they undertake, often lead to difficulties which result in human and financial loss. The history of these patients is in fact filled with losses in social standing, losses to family and friends, and bankruptcy. The choices these patients make are no longer personally advantageous and are socially inadequate. They are a far cry from the kind of choices these same human beings made prior to the onset of neurological disease and the appearance of a focal area of brain damage.

My colleagues and I have identified the brain area whose damage causes the problem (Damasio 1994). It is located in the prefrontal part of frontal lobe. We have found that, in order for this tragic condition to appear, one particular sector of the prefrontal lobe must be damaged: the ventral and medial sector. Moreover, when the condition conforms to the history I have just outlined, other territories of the prefrontal cortex must be largely spared. Knowing which part of the brain was related to the problem was an important advance. It helped us with the explanation of what went wrong and it opened a window onto the normal neurobiological mechanisms which were disrupted by the lesion. However, knowing *where* is not knowing *how*, and an adequate explanation of these patients' troubles requires a deeper understanding of the disordered process, at multiple levels, both neural and mental. If knowing where was clear, at the level of large-scale neural systems, knowing how was a complete mystery. This is why: while beset by faulty decision-making, these patients' intellect remains normal, as measured by conventional IQ tests, and so do the learning and retention of factual knowledge and the learning and retention of skills. The ability to use logic in the solution of problems commonly posed in neuropsychological testing is also normal. Basic attention, working-memory, and language are not affected. The ability to make estimates, the ability to judge recency and frequency of events, and performance in the Wisconsin Card Sorting Test (which has been thought to relate to frontal lobe function) are also normal. Given that impairments in such abilities and test performances have been seen as a hallmark of frontal lobe dysfunction, it is noteworthy that they are not compromised. Just as puzzling is the fact that these patients' repertoire of social knowledge is retained and can be accessed, in a laboratory situation. In brief, these patients' disturbance cannot be explained as the result of inadequate knowledge, or of a defect in intellectual ability, language, or basic working-memory and attention. The patients pose a double challenge. Although the disorder is blatant in its ecological niche, there has been no laboratory probe to detect it (let alone measure it), and no satisfactory account of the cognitive mechanisms and fine neural mechanisms underlying it. I am pleased to report some progress on both fronts. First I will describe the somatic marker hypothesis, with which I believe we can account for the condition. Then I will describe new laboratory probes designed to detect and measure aspects of the problem.

THE SOMATIC MARKER HYPOTHESIS

I developed the idea for the somatic marker hypothesis after realizing that the patients no longer had a normal ability to express emotion and experience feelings in situations in which both would have been expected. Based on reports from the patients' families such emotions and feelings had been present before the neurological illness set in. In other words, patients with ventromedial damage were intact in every neuropsychological test we tried and could thus be described as being of normal intellect. Yet their defective decision-making was not quite an isolated phenomenon: it had an accompaniment which consisted of abnormalities in emotion and feeling. In the absence of other impairments that might effectively account for the salient aspects of the condition, I then assumed that the defect in emotion and feeling along with its underlying neurobiological components, might be a plausible cause. In the paragraphs below I summarize my proposal (see Damasio 1994 for details).

First, some neural systems in the prefrontal cortex learn the association between certain classes of complex situation (the elements that compose the situation, the options available for response to it, the outcomes of engaging each and every option, and so on) and the type of emotional state usually associated with that class of situation in previous individual experience. I designated such emotional changes under the general term 'somatic state', because I see emotion as expressing itself most importantly (albeit not solely) through changes in body state, and because I believe that the results of emotion are primarily represented in the brain in the form of transient changes in the activity pattern of the brain regions which receive and map body signals (the somatosensory structures). Note that by somatic I refer to musculoskeletal, visceral, and internal milieu components of the soma and not just to the musculoskeletal aspect.

Second, when a situation of a given class recurs, the system in the prefrontal cortex which has previously acquired the link between the class of situation and the class of somatic state, that is the cerebral cortices in the ventromedial prefrontal region, can now trigger the re-activation of the appropriate somatosensory pattern. This re-activation can be carried out via a 'body loop', in which the soma actually changes in response to the activation, and the result of the ensuing changes is relayed to somatosensory cortices; or via an 'as-if' loop, in which the body is bypassed and the re-activation signals are conveyed directly to the somatosensory cortices which then adopt the appropriate pattern. The results of either mechanism may be overt, that is conscious, or covert, that is non-conscious.

Third, the establishment of a somatosensory pattern appropriately linked to the situation operates to qualify the factual evocations pertinent to the situation. You can imagine it as an adjective that marks the quality of the situation it refers to. It is through this qualifying action that the process of reasoning over multiple options and multiple future outcomes becomes constrained. When the image which results from the somatosensory pattern is juxtaposed to the images which describe a scenario of future outcome, the somatosensory pattern, positive or negative, marks the scenario as good or bad. When the process is overt, the somatic state can thus act as an alarm signal and lead to the rejection of an option/outcome pair, or as an incentive signal and lead to its endorsement. When it operates covertly the somatic state constitutes a biasing signal but still affects the activity of the network in the same manner. Finally, the somatosensory patterns also act as boosters to attention and working-memory.

In summary, the hypothesis suggests that somatic markers normally help to constrain the decision-making space by making that space manageable for logic-based, cost-benefit analyses. In situations in which there is remarkable uncertainty about the future and in which the decision can be influenced by previous individual experience, such constraints permit the organism to decide efficiently within short time intervals. In the absence of a somatic marker, options and outcomes are undifferentiated and logic must operate over too many option/outcome pairs and thus more slowly. Ventromedial frontal lobe patients often show this pattern of behaviour. In this text, I do not address the condition of other patients with frontal lobe damage whose lesions are located in other anatomical sectors and who have both defects in decision-making, and measurable defects in some of the abilities that are preserved in patients with ventromedial lesions. Patients with these other lesion patterns may or may not be accountable by the somatic marker hypothesis.

A NEURAL SYSTEM FOR SOMATIC MARKERS

Why is it that ventral and medial prefrontal cortices are critical to the type of personal and social reasoning and decision-making defects noted in our patients? These cortices, judging from what is known of non-human primate neuroanatomy, receive signals from all sensory modalities, directly or indirectly. In turn, they are the only known source of signals from frontal regions toward central autonomic control structures (those that control visceral activity). The source references for this neuroanatomical foundation include: Chavis and Pandya 1976; Jones and Powell 1970; Pandya and Kuypers 1969; Nauta 1971; Amaral and Price 1984; Goldman-Rakic *et al.* 1984; Porrino *et al.* 1981; Van Hoesen *et al.* 1972, 1975.

This anatomical design is compatible with the role I propose for the ventromedial cortices. I believe these cortices contain 'convergence zones' which hold a record of temporal conjunctions of activity in varied neural regions (for example sensory cortices, limbic structures) hailing from both external and internal stimuli. (For a review of the convergence zone framework see Damasio 1989 *a*, *b*; Damasio and Damasio 1994.) This would be a record of signals from regions that were active simultaneously and which, as a set, define a given situation. A critical output of the ventromedial convergence zones would be to autonomic effectors. In neural terms, the system would operate as follows: when parts of certain exteroceptive-interoceptive conjunctions are re-processed, consciously or not, their activation is signalled to ventromedial cortices, which in turn activate somatic effectors (in amygdala, hypothalamus, and brainstem nuclei). One might say that this process is an attempt to reconstitute the kind of somatic state that belonged to the conjunction. The re-enacted somatic state can then be signalled to cortical and subcortical somatosensory processing structures, and either be consciously perceived or trigger a covert process modifying appetitive or aversive behaviours. Ventromedial frontal damage would preclude this chain of events.

In other words, ventral and medial prefrontal cortices are the natural recipients of signals concerning the multiple-site representation of scenarios of complex situations, in several early sensory cortices. Moreover, ventromedial cortices are also recipients of signals from somatosensory structures and bioregulatory structures. Finally, ventromedial frontal structures can originate signals to bioregulatory structures such as the amygdala, hypothalamus, and brainstem.

The neural system network necessary for somatic markers to operate thus includes:

(1) Ventromedial frontal cortices which contain convergence zones that record links between (a) the dispositions that represent categorizations of certain complex situations and (b) the dispositions that represent the somatic states that have been prevalently associated with the situations referred to above;
(2) central autonomic effectors, for example the amygdala, which can activate somatic responses in viscera, vascular bed, endocrine system, and non-specific neurotransmitter systems; and
(3) somatosensory cortices and their interlocking projections (especially in the non-dominant hemisphere).

The evocation of a somatic marker for simple stimuli such as a sudden noise or flash of light requires a simpler system that can cope with behaviourally relevant stimuli that do not need the kind of complex information processing that social configurations require. The alternate system probably bypasses the cerebral cortex altogether and activates autonomic centres (for example amygdala and others) directly from the thalamus.

THE ROLE OF EMOTION

Why are somatic signals so critical to the process of reasoning and decision-making? My answer is that certain classes of situation, namely those that concern personal and social matters, are frequently linked to punishment and reward, and thus to pain, pleasure, and to the regulation of homeostatic states, including the part of the regulation that is expressed by emotion and feeling. It is not possible to represent neurally any of these aspects of cognition and biology without involving the somatosensory system.

A more general question is why a signal is needed at all, external to the images over which our reasoning is naturally conducted and which can represent all the logical manipulations one usually deems necessary and sufficient for considering an issue and deciding accordingly. The answer is that the uncertainty of outcomes, and the large dimension of the logical operations required to decide under uncertainty, make it advantageous to have a signal which helps constrain the decision-making space.

I suspect that the mechanism which the brain has used to select good responses from bad ones, in social situations, has evolved from a mechanism co-opted for behavioural guidance that was not originally devoted to the social affairs it has ultimately been co-opted for. My argument here is that nature would have evolved a highly successful mechanism of guidance to cope with problems whose answer might maximize survival by leading away from danger. The use of those mechanisms was later extended to the social realm. Likewise, it is plausible that a system geared to produce markers and signposts to guide 'social' responses, would have been further adapted to assist with 'intellectual' decision-making over other problems which are indirectly linked to precisely the same framework of survival versus danger, of ultimate advantage versus disadvantage, of ultimate gain and balance versus loss and disequilibrium. The somatic markers would act covertly to highlight, in the form of an attentional mechanism, certain components over others, and to direct, in effect, the go, stop, and turn signals necessary for much decision-making and planning on even the most abstract of topics. On this issue, Shallice has also proposed that some form of marker is needed in decision-making, although he has not specified the neurobiological nature of his marker and it may well be different from the device I am proposing here (Shallice and Burgess 1993).

In conclusion, in normal individuals, certain situations require high-order composite memories formed by 'facts' and by the 'body states' which usually accompany those facts in an individual's experience. The 'fact' memories are held in dispositional form in the appropriate association cortices. The 'body state' memories do not need to be held permanently, since body states can be re-enacted on demand. Only the memory of the *linkage* between certain classes of situation and certain body states must be held permanently, and I believe the requisite system is in ventral and medial prefrontal cortices.

Patients with ventromedial frontal lobe damage fail to evoke part of the composite memory for a class of situation – the part that describes the association between the class of situation and the somatosensory state linked to the situation. The factual knowledge component of the composite memory can still be evoked, but somatic states cannot be re-enacted, overtly or covertly, relative to those facts. This limitation poses no problem for situations that have minimal somatic state associations in previous experience, but is catastrophic for situations that do (for additional details on emotion see Damasio 1994, 1995).

TESTING THE SOMATIC MARKER HYPOTHESIS

The gambling experiments

One approach to the testing of the somatic marker hypothesis relied on a novel card gambling task. As described elsewhere (Bechara *et al.* 1994) the task is an attempt to create, in the laboratory, a realistic situation in which subjects gradually learn how to play a card game, to their best advantage, in situations of limited knowledge about the rules, and under the control of rewards and penalties. The task operates as follows: the people sit in front of four decks of cards equal in appearance and size, and are given a $2000 loan of play money (facsimile US bills). They are told that the game requires a series of card selections, one card at a time, from any of the four decks, until they are told to stop. They are also told that (1) the goal of the task is to maximize profit on the loan of play money, (2) they are free to switch from any deck to another, at any time, and as often as wished; but (3) they are not told ahead of time how many card selections must be made; the task is stopped after 100 card selections. After *each* card turning, they receive some money. The amount is announced after the turning and varies with the deck.

Turning any card from deck A or deck B yields $100; turning any card from deck C or deck D yields $50. After turning *some* cards of any deck, however, the subjects are *both* given money *and* asked to pay a penalty. Again the amount is announced after the card is turned and varies with the deck and the position in the deck according to a schedule unknown to the subjects. The ultimate yield of each deck varies because the penalty amounts are higher in the high-paying decks (A and B), and lower in the low-paying decks (C and D). For example, after turning 10 cards from deck A, the subjects have earned $1000, but they have also encountered 5 unexpected punishments bringing their total cost to $1250, and incurring a net loss of $250. They encounter the same problem on deck B. On the other hand, after turning 10 cards from decks C or D, the subjects earn $500, but their unpredicted punishments only amount to $250, that is subject nets $250. In short, decks A and B are equivalent in terms of overall net loss over the trials. The difference is that in deck A, the punishment is more frequent, but of smaller magnitude, whereas in deck B, the punishment is less frequent but of higher magnitude. Decks C and D are also equivalent in terms of overall net loss. In deck C, the punishment is more frequent and of smaller magnitude, while in deck D the punishment is less frequent but of higher magnitude. Decks A and B are 'disadvantageous' because they cost the most in the long run, while decks C and D are 'advantageous' because they result in an overall gain in the long run.

The performances of normal, control group (21 women and 23 men) in this task were compared to those of a ventromedial prefrontal group (4 men and 2 women). The age range of normal controls was from 20 to 79 years; for ventromedial people it was from 43 to 84 years. About half the people in each group had secondary education, and the other half had tertiary education.

The results were strikingly clear cut. Normal people and patients without frontal damage sample from all decks for a while and gradually begin playing more frequently from the good deck than from the bad. About half way through the game they finally adopt this strategy, never abandon it, and win. Ventromedial frontal lobe patients, on the contrary, continue to play predominantly from the bad decks, in spite of repeated losses. As a result they lose all of their loan money and need to borrow more.

Although the gambling task involves a long series of gains and losses, it is not possible for people to perform an exact calculation of the net gains or losses generated from each deck as they play. For instance, a normal control group with superior memory and IQ, whom we asked to think aloud while performing the task and keep track of the magnitudes and frequencies of the various punishments, could not provide figures for the net gains or losses from each deck. The subjects must rely on their ability 'to sense' which decks are risky and which are profitable. The performance profile of ventromedial patients is comparable to their real-life inability to decide advantageously, especially in personal and social matters, a domain for which in life, as in the task, an exact calculation of the future outcomes is not possible and choices must be based on approximations. This task offers, for the first time, the possibility of detecting and measuring these patients' elusive impairment in the laboratory.

In previous discussions of these results (Bechara *et al.* 1994; Damasio 1994), we considered three possibilities for why the target patients make choices that have high immediate reward but severe delayed punishment. The first is that patients are so sensitive to reward that the prospect of future (delayed) punishment is outweighed by that of immediate gain. The second is that they are insensitive to punishment, and thus the prospect of reward always prevails, even if they are not abnormally sensitive to reward. The third is that they are generally insensitive to future consequences, positive or negative, and thus their behaviour is mostly guided by immediate prospects. Subsequent work with variants of these tasks (to be published) suggest that neither insensitivity to punishment nor hypersensitivity to reward account for the defect. Rather the patients seem unresponsive to future consequences, whatever they are.

Evidence from other studies indicates that these patients retain and access the knowledge necessary to conjure up options of actions and scenarios of future outcomes. What they fail to do is act on such knowledge. Elsewhere we have proposed several possible mechanisms of defect (Damasio 1994; Bechara *et al.* 1994). For instance, it is possible that the representations of future outcomes that these patients evoke are unstable, that is they are not held in working memory long enough for attention to enhance them and for reasoning strategies to operate on them. This mechanism invokes a defect along the lines proposed for behavioural domains dependent on dorsolateral prefrontal cortex networks (Goldman-Rakic 1987). Defects in temporal integration and attention would be related to this account (Fuster 1989; Posner and Petersen 1990). Another possibility is that the representations of future outcomes might be stable but would not be marked with a negative or positive value, the value mark for a cognitive scenario would be missing (Damasio 1994; Damasio *et al.* 1991).

Psychophysiological correlates of the gambling performance

In a further test to the somatic marker test hypothesis my colleagues and I undertook a continuous monitoring of skin conductant responses, while normal people and patients were engaged in the gambling task (Bechara *et al.* 1996). We already knew from previous experiments that skin conductance responses to cognitive processing could be altered in these patients (Damasio *et al.* 1991; Tranel 1994; Tranel *et al.* 1995) and we were not disappointed. The most salient result of this study was the finding that during the time window that precedes the selection of a card from a given deck normal people begin to respond with high amplitude skin conductance responses (SCRs), whenever they are about to make a selection from a bad deck. They show no comparable responses when they are about to make a selection from a good deck. As the task unfolds, the SCRs to the bad decks continue to appear systematically and they rise in amplitude. This does not happen for the responses associated with the good decks. Quite remarkably, no such anticipatory responses are seen in the patients with ventromedial frontal damage at all.

One possible interpretation is that the SCR is part of a very early and automated alarm signal, which is triggered, as proposed in the somatic marker hypothesis, from the ventromedial region. The signal affects further processing of the factual knowledge connected with the situation by marking a particular option–outcome pair with a negative bias. Incidentally, this interpretation holds whether the signal is overt and fully appreciated in consciousness, or covert and entirely operated at an unconscious level.

An alternative interpretation is that normal people reason, early on, that certain decks are bad and certain decks are good, and that, on the basis of their cognition of 'badness' and 'goodness', they generate a somatic response which is indexed by the SCR. For reasons that have to do with my perspective on the evolutionary biology and adaptive value of an automated somatic marker device, I find the latter interpretation less plausible. Also, recent studies in our laboratory suggest that normal people begin producing their SCRs to bad decks long before they have, according to their testimony, any notion whatsoever of the good or evil nature of each deck and of the design of the game they are playing.

CONCLUDING REMARKS

Rationality, the process of reasoning and decision-making in a personal and social domain, depends on the prefrontal cortex and, specifically, on the ventromedial sector of these cortex. Overall, the operation depends on a large-scale neural system with many components, which work in concert with the ventromedial sector. Several key components of the system are structures which regulate life-maintenance by the process of homeostasis – the collection of actions which permits the organism's internal environment to achieve physiological balances. We are largely unaware of this regulatory process. Part of what little we come to know of its doings is expressed in the form of emotions which, by and large, play themselves out in the theatre of our body. Since I view body (somatic) changes as the essential aspect of the process of emotion, albeit not the only one, I have called the hypothesis that subsumes my ideas on the neurobiology of rationality the 'somatic marker hypothesis'.

The connections that I am establishing between emotion and reason attempt to restore emotion to a central role in the understanding of cognition. The least one might say about this attempt is that it is long overdue. Towards the end of the nineteenth century the understanding of emotion did flourish – Charles Darwin placed emotion in an evolutionary perspective and made a landmark contribution to the biology of emotion; William James advanced a revolutionary hypothesis concerning the process of emotion; and the much maligned Sigmund Freud concerned himself with the role of emotion in the development of abnormal mental states. Strangely, however, twentieth century neuroscience and even cognitive science turned their backs on emotion, the notable exceptions merely serving to confirm the truth of this general statement. It is perhaps no coincidence that ours is also the century during which rationality has been increasingly seen as an exercise in logic, and, in its second half, as an algorithmic process best described in terms of the ever more popular computer metaphor for the human brain/mind.

My hope is that the sort of work being pursued, not only in my laboratory but in others, may help the elucidation of some of the biological mechanisms of emotion and reason, and, no less importantly, may help others see emotion not as the evil twin of reason, but rather as a very natural and inextricable component of the nature of being rational, for better and for worse. It is not the case that we should reason with emotion alone but rather that we cannot reason without it. Of course, I would also hope that this would make an old friend, David Hume, smile most warmly.

In closing I must return to the issues of environment and of the future of human life. On this matter, let me call attention to the striking continuity between three distinct kinds of environment we may consider. First, the *internal environment* of living organisms (Claude Bernard's internal milieu), on the delicate balance of which life hangs moment by moment. Second, the *physical environment* which surrounds us, and the regulation of which is also a delicate enterprise at cosmic scale. Third, the *cultural environment* which the self-conscious minds of living beings have been producing in a social setting. It is apparent that nature, through the thick and thin of evolution, has developed mechanisms to regulate the internal environment of living beings and that those mechanisms largely operate automatically and very efficiently. Nature has also developed a way to cope with the large scale enterprise of regulating the global physical environment with very little help and often much harm from the self-conscious living beings that use it. What nature has not yet developed is an effective means to regulate a survivable cultural environment at the global scale. One might say that, as we approach the millennium, we are still witnessing the natural evolution of the regulatory devices for this third kind of environment. Since those of us alive today had no opportunity to witness significant evolutionary changes in the physical and biological environments, we may regard the chance of witnessing social and cultural evolution as a privilege. Or perhaps not, given that what we have to witness is often not pleasant and sometimes dangerous. Pleasant or not, the opportunity is exciting especially when one realizes that, in some part, the result of that evolution is under our control, and that part of that control depends on the proper understanding of how human beings emote and reason, for better and for worse.

ACKNOWLEDGEMENT

This research was supported in part by NIH NINDS P01 NS 19632, and the Mathers Foundation.

REFERENCES

Amaral, D.G. and Price, J.L. (1984). Amygdalo-cortical projections in the monkey (Macaca fascicularis). *J. Comp. Neurol.*, **230** 465–96.

Bechara, A., Damasio, A.R., Damasio, H., and Anderson, S.W. (1994). Insensitivity to future consequences following damage to human prefrontal cortex. *Cognition*, **50**, 7–12.

Bechara, A., Tranel, D., Damasio, H., and Damasio, A.R. (1996) Failure to respond autonomically to anticipated future outcomes following damage to prefrontal cortex. *Cerebral Cortex*, **275**, 1293–4.

Chavis, D.A. and Pandya, D.N. (1976). Further observations on corticofrontal connections in the rhesus monkey. *Brain Res.*, **117**, 369–86.

Damasio, A.R. (1979). The frontal lobes. In *Clinical neuropsychology* (ed. K.M. Heilman and E. Valenstein), pp.360–412. Oxford University Press, New York.

Damasio, A.R. (1989a). The brain binds entities and events by multiregional activation from convergence zones. *Neural Computation*, **1**, 123–32.

Damasio, A.R. (1989b). Time-locked multiregional retroactivation: a systems level proposal for the neural substrates of recall and recognition. *Cognition*, **33**, 25–62.

Damasio, A.R. (1994). *Descartes' error: emotion, reason, and the human brain.* Grosset/Putnam, New York; also 1995, Picador/Macmillan, London.

Damasio, A.R. (1995). Toward a neurobiology of emotion and feeling: operational concepts and hypotheses. *The Neuroscientist*, **1**, 19–25.

Damasio, A.R. and Damasio, H. (1994). Cortical systems for retrieval of concrete knowledge: the convergence zone framework. In *Large-scale neuronal theories of the brain* (ed. C. Koch), pp. 61–74. MIT Press, Cambridge, MA.

Damasio, A.R., Tranel, D., and Damasio, H. (1991). Somatic markers and the guidance of behavior: theory and preliminary testing. In *Frontal lobe function and dysfunction* (ed. H.S. Levin, H.M. Eisenberg, and A.L. Benton), pp.217–29. Oxford University Press, New York.

Fuster, J.M. (1989). *The prefrontal cortex: anatomy, physiology, and neuropsychology of the frontal lobe* (2nd ed). Raven Press, New York.

Goldman-Rakic, P. (1987). Circuitry of primate prefrontal cortex and regulation of behavior by representational memory. In *Handbook of physiology: the nervous system* (ed. F. Plum and V. Mountcastle) Vol. 5, pp.373–401. American Physiological Society, Bethesda, MD.

Goldman-Rakic, P.S., Selemon, L.D., and Schwartz, M.L. (1984). Dual pathways connecting the dorsolateral prefrontal cortex with the hippocampal formation and parahippocampal cortex in the rhesus monkey. *Neuroscience*, **12**, 719–43.

Jones, E.G. and Powell, T.P.S. (1970). An anatomical study of converging sensory pathways within the cerebral cortex of the monkey. *Brain*, **93**, 793–820.

Nauta, W.J.H. (1971). The problem of the frontal lobe: a reinterpretation. *J. Psychiatr. Res.*, **8**, 167–87.

Pandya, D.N. and Kuypers, H.G.J.M. (1969). Cortico-cortical connections in the rhesus monkey. *Brain Res.*, **13**, 13–36.

Porrino, L.J., Crane, A.M., and Goldman-Rakic, P.S. (1981). Direct and indirect pathways from the amygdala to the frontal lobe in rhesus monkeys. *J. Comp. Neurol.*, **198**, 121–36.

Posner, M.I. and Petersen, S.E., (1990). The attention system of the human brain. *Ann. Rev. Neurosci.*, **13**, 25–42.

Shallice, T. and Burgess, P.W. (1993). Supervisory control of action and thought selection. In *Attention: selection, awareness, and control: a tribute to Donald Broadbent* (ed. A. Baddeley and L. Weiskrantz), pp.171–87. Clarendon Press, Oxford.

Tranel, D. (1994). 'Acquired sociopathy': the development of sociopathic behavior following focal brain damage. In *Progress in experimental personality and psychopathology research* (ed. D.C. Fowles, P. Sutker, and S.H. Goodman.) Vol. **17**, pp.285–311. Springer, New York.

Tranel, D., Damasio, H., and Damasio, A.R. (1995). Double dissociation between overt and covert face recognition. *J. Cog. Neurosci.* (in press).

Van Hoesen, G.W., Pandya, D.N., and Butters, N. (1972). Cortical afferents to the entorhinal cortex of the rhesus monkey. *Science*. **175**, 1471–3.

Van Hoesen, G.W., Pandya, D.N., and Butters, N. (1975). Some connections of the entorhinal (area 28) and perirhinal (area 35) cortices of the rhesus monkey: II. Frontal lobe afferents. *Brain Res.*, **95**, 25–38.

4
Perceiving social and physical environments

Andy Young

Professor Andy Young, BSc, PhD, DSc, holds a special appointment at the Medical Research Council's (MRC) Applied Psychology Unit at Cambridge and is an Honorary Professor in the School of Psychology at the University of Wales, Cardiff. He took his first degree, in psychology, at the University of London in 1971 and then undertook research for his doctorate at the University of Warwick. He held a Lectureship in Psychology at the University of Aberdeen and then, from 1976 to 1989, a Lectureship and Readership at the University of Lancaster. In 1989 he was appointed Professor of Psychology at the University of Durham, a post which he held for four years until taking up his present post at Cambridge. In 1994 Professor Young was co-recipient of the British Psychological Society's Cognitive Psychology Award; in 1995 the Society awarded him its President's Award for his distinguished contribution to psychological knowledge. He is a leading expert on impairments of face recognition caused by brain injury; he is the author of a large number of publications on this topic and on other aspects of neuropsychology.

INTRODUCTION

The goal of perception is to create representations of external events in ways that permit effective actions in the world that an organism inhabits. As human beings, we therefore need not only to represent the physical environment but also the identities, emotions, and intentions of other people. Sometimes this process goes wrong; temporarily when we suffer mental lapses, or for prolonged periods due to disease or brain injury. The resulting perceptual failures offer unique insights into how the brain creates its conscious and non-conscious representations of our social and physical worlds.

We will look at findings about mind, brain, and environment from neuropsychology, which involves studying impairments caused by brain injury. As we will see, some of these impairments can be very circumscribed, and these have direct implications for understanding how the brain analyses the environment, and how some types of mental experience relate to underlying brain mechanisms. Face perception will be used as an especially instructive example, where alternative accounts of mental organization can be compared, contrasted, and teased apart.

A good illustration of selective impairments of mental abilities after brain injury comes from an important finding made by Newcombe and Russell (1969). They were studying a group of ex-servicemen who had suffered penetrating wounds to the brain, mostly from shrapnel injuries during the Normandy landings and the subsequent liberation of Europe. In the 1960s, when the tests reported by Newcombe and Russell (1969) were carried out, the ex-servicemen had lived with these injuries for some 20 years; more than enough time for any recovery, relearning, or rehabilitation. Mostly, they were leading perfectly normal lives. Yet they could still show striking problems on certain tests.

Newcombe and Russell (1969) used two tasks, which are illustrated in Fig 4.1; these were a perceptual task which involved determining the age and sex of contrast-enhanced faces, and a visually-guided maze learning task. Many of the ex-servicemen with right hemisphere lesions were found to be impaired on these tasks, yet Newcombe and Russell (1969) noted that there was no overlap between the scores of the men who were most severely impaired in face perception or most severely impaired in maze learning. Instead, some ex-servicemen were very poor at the face perception task yet were able to learn the visual maze without difficulty; these were found to have lesions involving posterior parts of the right temporal lobe. Other cases with the opposite pattern of impaired maze learning and normal face perception had high right posterior parietal injuries.

A later report on two contrasting cases of visual perceptual and spatial impairment, with autopsy findings, confirmed these conclusions (Newcombe *et al.* 1987). Fig. 4.2 shows the performance of these two ex-servicemen (VRD and JS) in relation to comparison groups of men with left hemisphere injuries or neurologically normal controls. Performance is expressed as a *z*-score, indicating the number of standard deviations by which each person differs from control performance on each task. The standard deviation is a measure of the variability of the scores of the control subjects; the higher the *z*-score for a given task, the greater is the degree of difference

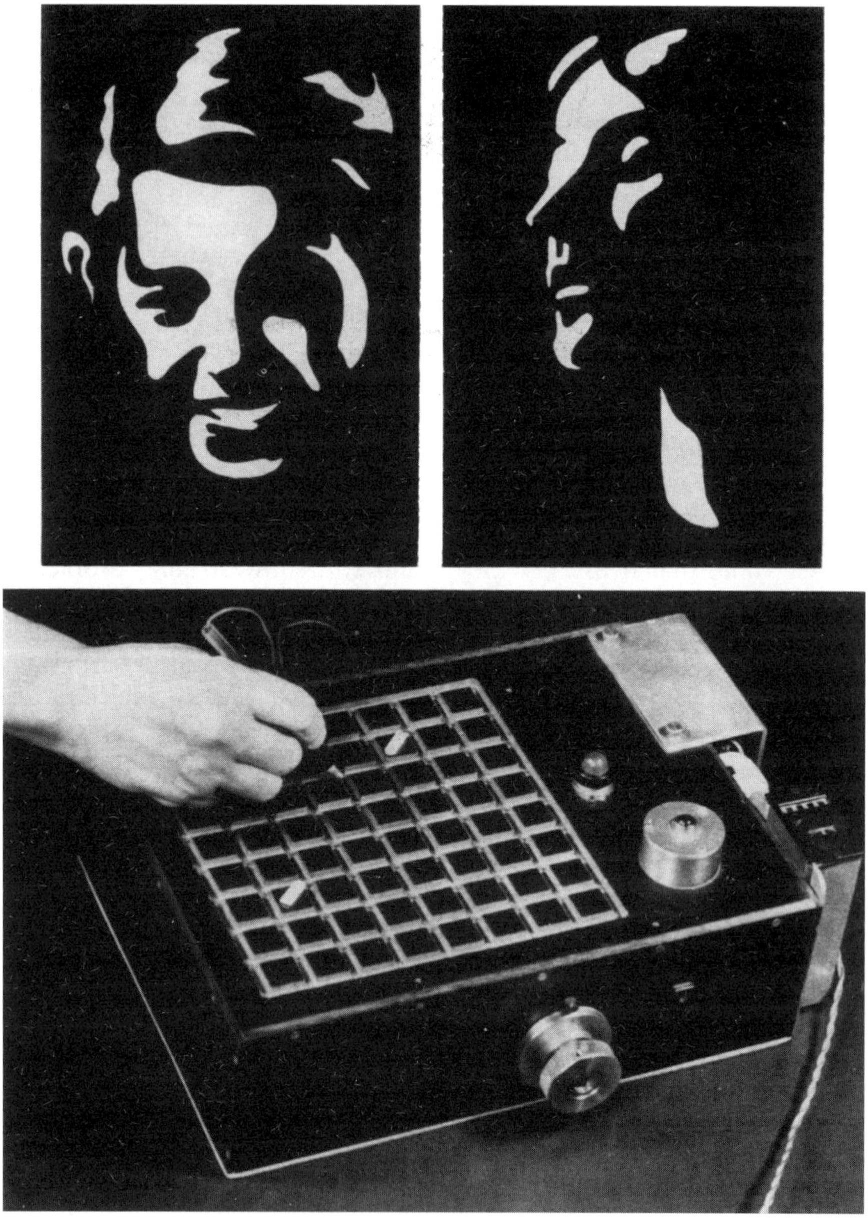

Fig. 4.1 Examples of high-contrast faces and stylus maze used by Newcombe and Russell (1969). The faces must be classified as man, woman, old man, old woman, boy, or girl. The maze requires finding a path between the two arrows; errors are signalled by a red light.

from controls (that is in Fig. 4.2 higher *z*-scores represent more impairment). Both VRD and JS remained of normal intelligence (as estimated from their vocabulary) yet VRD, who had sustained right parietal lobe injury, had great difficulty learning the visual maze and JS, who had sustained right temporal lobe injury, was very poor at the face perception (closure) task. In line with these findings, Newcombe *et al.* (1987) noted that JS was also poor at recognizing familiar faces.

In neuropsychology, this kind of pattern is called a double dissociation; each person is bad at what the other person is good at. Double dissociations are important because they point clearly to the possibility that different brain regions underlie the different abilities, allowing one to be impaired whilst the other remains relatively intact. This inference is especially compelling from the data of Newcombe *et al.* (1987) because despite being so poor at maze learning VRD had no problem at all with the face perception task, whereas JS made many errors in face perception but experienced not the slightest difficulty in learning the maze.

The significance of this double dissociation of what Newcombe and Russell (1969) called visual perceptual and spatial deficits has been enhanced by subsequent advances in our understanding of the neurophysiology of the visual system (Cowey 1994). These have shown that the brain regions involved in vision are much more extensive than the classical visual cortex (in the occipital lobe), and that there seem to be many discrete visual areas. These visual areas can be grouped into separate streams projecting from occipital through temporal lobes, and from occipital through parietal lobes; Ungerleider and Mishkin (1982) suggested that these can be considered as partially independent systems, with the occipito-temporal stream responsible for interpreting *what* an object is, and the occipito-parietal stream determining *where* it is located. These neurophysiological findings and theories are exactly in line with the observations reported by Newcombe and her colleagues (Newcombe *et al.* 1987; Newcombe and Russell 1969).

Other examples of selective perceptual impairment can be considered to go beyond Ungerleider and Mishkin's (1982) 'what versus where' distinction, because this characterization misses something important about perceptual mechanisms; they are also geared to interpreting the social as well as the physical environment. Much of the 'what' system seems to be devoted to this purpose. We will look at some of the types of deficit which can affect social perception after brain injury, concentrating on the perception of different social signals

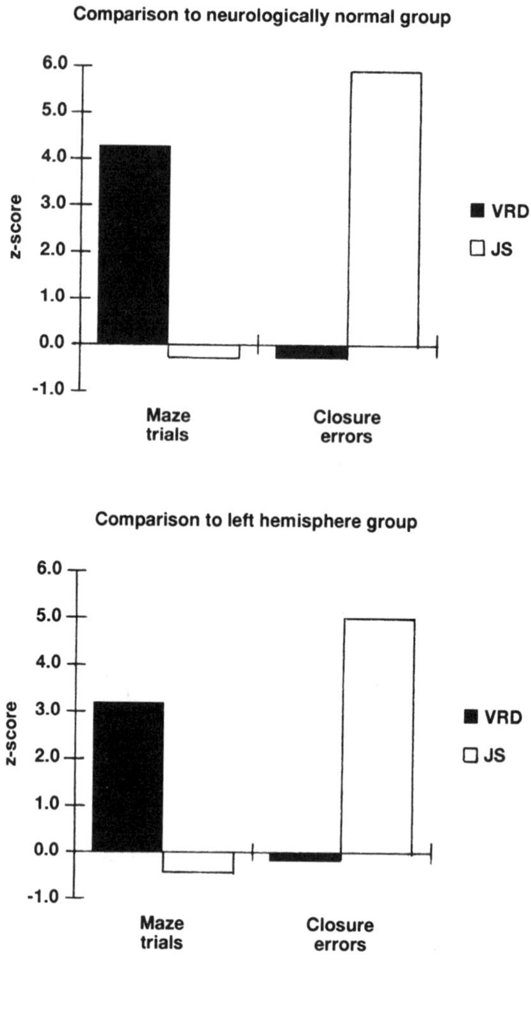

Fig. 4.2 Contrasting cases (VRD and JS) of visual perceptual and spatial impairments reported by Newcombe *et al.* (1987). Performance is expressed here as *z*-scores with respect to comparison groups who were neurologically normal or who had unilateral injuries of the left cerebral hemisphere. VRD needed many trials to learn the maze, but had no problem with the face perception (closure) task. JS made many errors in face perception (closure), but experienced no difficulty in learning the maze.

given by the face, and exploring their implications for our perception of the social and physical worlds. We will consider impairments affecting the recognition of identity and emotion, and examine some of the differences between conscious and non-conscious representations of social and physical environments, to see what these imply about the functions of consciousness.

DISSOCIABLE IMPAIRMENTS OF FACE PROCESSING ABILITIES

We begin by looking more carefully at problems akin to those of JS (see Fig. 4.2), who was poor at the face perception task shown in Fig. 4.1. Quite reasonably, Newcombe and Russell (1969) had interpreted this as likely to reflect a general problem of visual 'closure', rather than a problem related to perceiving faces *per se.* However, subsequent research has revealed a more complex and interesting picture.

One obvious possibility to explore is that people like JS have a basic perceptual difficulty which interferes with their ability to see faces properly. A specific version of this idea had been suggested by Sergent (1983, 1984), who considered that the left and right cerebral hemispheres were differentially sensitive to information carried by different spatial frequencies. To investigate this, Newcombe *et al.* (1989) measured the spatial contrast sensitivity functions of neurologically normal controls and groups of ex-servicemen with unilateral brain lesions affecting the left or right cerebral hemispheres. The idea underlying the spatial contrast sensitivity function will not be familiar to all readers; in essence, all visual stimuli can be decomposed mathematically into signals varying between high spatial frequencies (fine detail, changing rapidly across small areas of the scene) and low spatial frequencies (coarse detail, relatively unchanging across broad areas). This is important because the human visual system is naturally more sensitive to information carried by intermediate than by high or low spatial frequencies. The spatial contrast sensitivity function charts this by measuring the lowest contrast at which signals at different spatial frequencies can be detected.

Fig. 4.3 shows mean spatial contrast sensitivity functions for ex-servicemen with unilateral injuries of the left or right cerebral hemispheres, and a neurologically normal control group matched for age and intelligence (Newcombe *et al.* 1989). The characteristic inverted U-shaped function, with maximum sensitivity to intermediate spatial frequencies, is clearly seen in the control data.

The ex-servicemen performed more poorly than controls, but the impairment was constant across all spatial frequencies and did not differ between left or right hemisphere injured groups. In summary, people with unilateral brain injuries do have somewhat impaired vision, but there is no hint of any differential deficit between those with left or with right hemisphere injuries.

This pattern makes an instructive contrast with problems in face perception measured for the same subject groups. To do this, Newcombe *et al.* (1989) used a 'face decision' task, in which people were asked to decide as quickly as possible whether or not a line drawing was of a properly organized or slightly scrambled face; examples of the stimuli are shown in Fig. 4.4, and mean reaction times are presented in Fig. 4.5. The ex-servicemen with right hemisphere lesions show the longest reaction times, indicating that they had problems in carrying out this simple task. The interest of the contrast between the data presented in Fig. 4.3 and Fig. 4.5 is that although both groups of ex-servicemen showed equally impaired

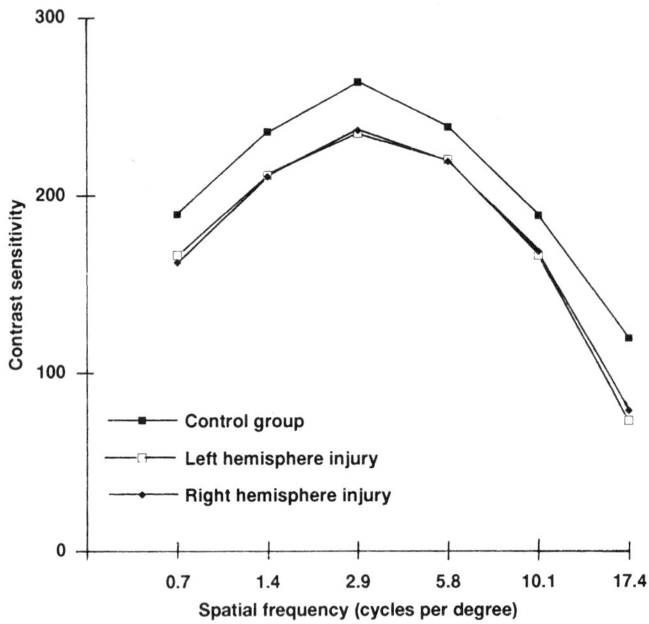

Fig. 4.3 Mean spatial contrast sensitivity functions for ex-servicemen with unilateral injuries of the left or right cerebral hemispheres, and a neurologically normal control group (Newcombe *et al.* 1989).

vision (Fig. 4.3), the group with right hemisphere injuries were more severely impaired at perceiving faces (Fig. 4.5). This suggests that the right hemisphere contributes something to face perception over and above its involvement in more basic visual abilities.

Of course, one might argue that the data shown in Fig. 4.5 could actually reflect a rather different problem for the people with right hemisphere injuries. The most obvious possibility is that they are simply slower at performing any speeded task. However, Newcombe *et al.* (1989) carried out a further experiment which involved deciding whether photographs showed familiar (famous) or unfamiliar faces, and deciding whether printed names belonged to familiar or unfamiliar people. Mean reaction times are shown in Fig. 4.6. Again, the ex-servicemen with right hemisphere injuries performed most poorly when faces were used as stimuli, but when names were used it was the group with left hemisphere injuries who were now slowest. Hence it is evident that the perceptual problems of the right hemisphere group reflect something which is needed for the analysis of faces and not needed for words.

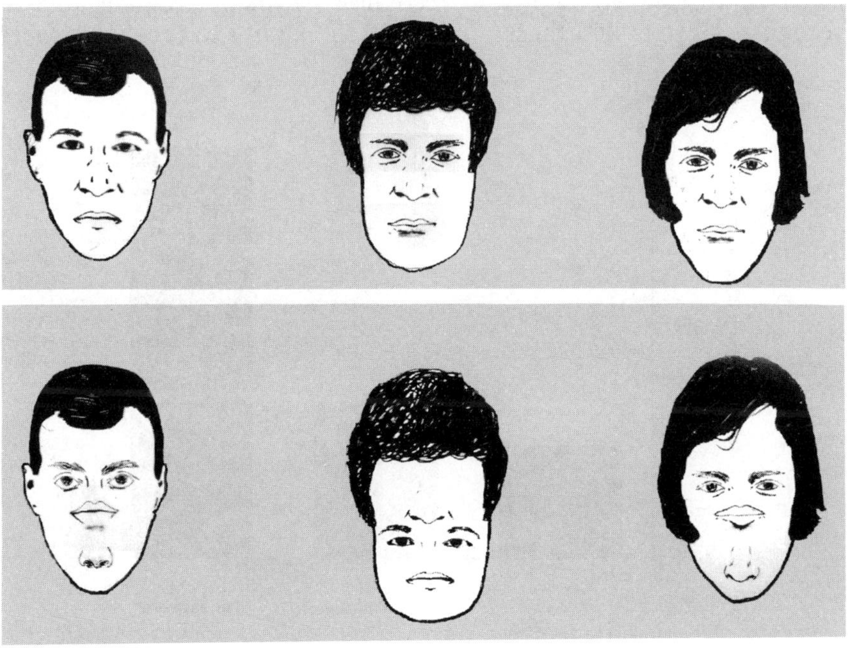

Fig. 4.4 Examples of stimuli from face decision task used by Newcombe *et al.* (1989). The task involves deciding as quickly as possible whether each is a properly organized or scrambled face.

In fact, the neuropsychological literature has often linked word recognition to the left cerebral hemisphere and face recognition to the right hemisphere, both for people with brain injuries and in studies of the neurologically normal brain (Ellis 1983; Young 1987). In addition, the face perception tasks used by Newcombe *et al.* (1989) were chosen on the basis that they were already known to be associated with right hemisphere superiority in normal perceivers (Young *et al.* 1985*a*, *b*).

What remains less clear is *why* word recognition should involve the left hemisphere and face recognition the right hemisphere, and especially *what* the cerebral hemispheres contribute to each ability. A particularly neat hypothesis has been suggested by Farah (1991). Farah was interested in the question of whether different neural systems might be needed to recognize words, faces, and everyday objects. To this end, she reviewed the literature on neuropsychological impairments of face, word, and object recognition, examining the different patterns of deficit which have been reported. This literature search showed that whereas people with brain injuries have been reported to lose the ability to recognize words yet remain able to recognize objects and faces, or to lose the ability to recognize faces

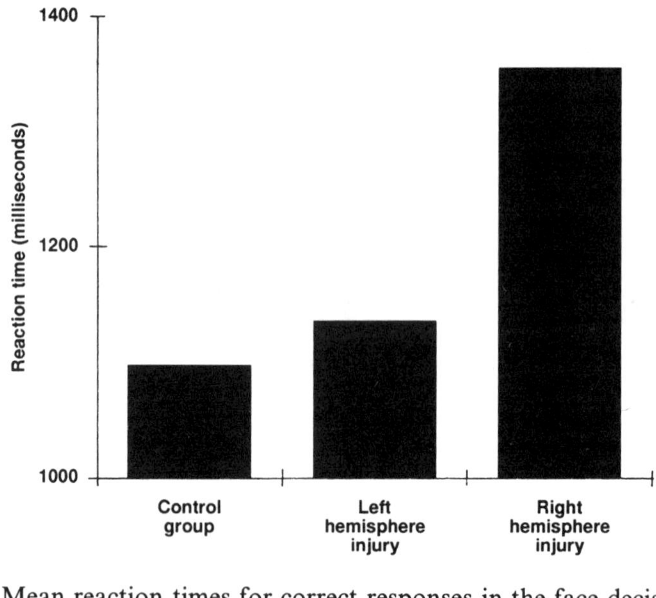

Fig. 4.5 Mean reaction times for correct responses in the face decision task for neurologically normal control subjects and ex-servicemen with unilateral injuries of the left or right cerebral hemispheres (Newcombe *et al.* 1989).

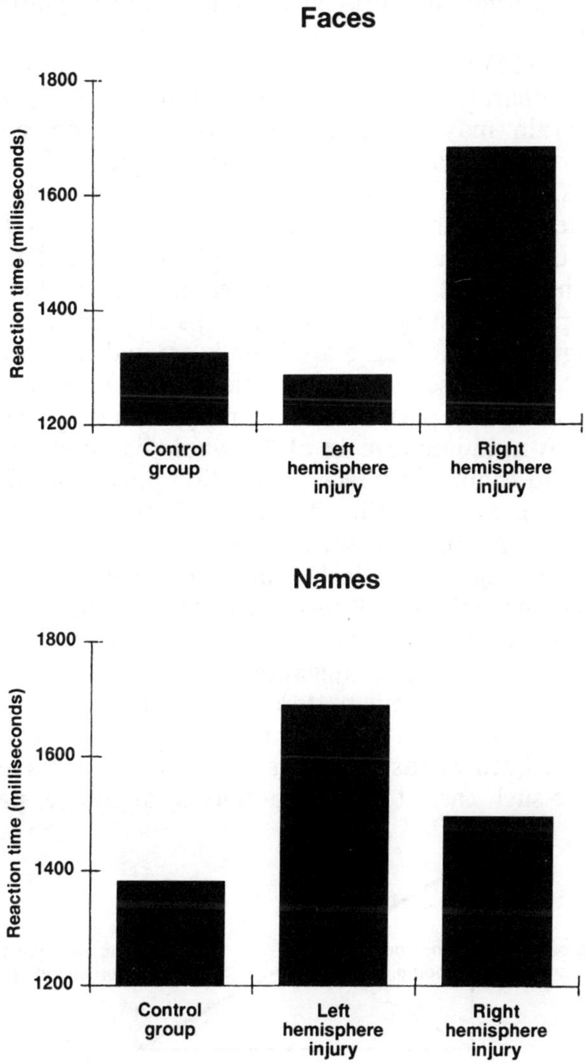

Fig. 4.6 Mean reaction times for correct responses in deciding whether faces or names are familiar, for neurologically normal control subjects and ex-servicemen with unilateral injuries of the left or right cerebral hemispheres (Newcombe *et al.* 1989).

whilst remaining able to recognize objects and words, there were at the time no convincing cases in which object recognition was compromised *without* impairment of *either* face recognition *or* word recognition.

As Farah (1991) pointed out, this suggests that, rather than there being three separate recognition systems for objects, words, and faces, the brain may make use of two underlying systems; one which is particularly appropriate for words but is also needed for some objects, and one which is particularly appropriate for faces but is also needed for some objects, as shown in Fig. 4.7. A plausible account would be that one system might be needed to represent stimuli (such as faces) which are not readily identifiable from their constituent parts (most eyes, noses, mouths are hard to recognize as isolated fragments) and hence are usually treated holistically, whilst another system represents stimuli which are readily decomposable into simpler nameable parts used to identify them (for example words into letters). A mild impairment of the whole-based encoding system could then lead to impaired face recognition only, since faces are the stimuli which place the greatest demands on this form of recognition, but more severe impairments would begin to compromise recognition of at least some objects. Similarly, mild impairment of the part-based encoding system would initially affect word recognition only, and then increasingly severe impairments would start to involve objects.

Although attractive and ingenious, Farah's (1991) hypothesis clearly predicts that certain patterns of deficit should never occur; the main prediction is that impaired recognition of objects with intact recognition of both words and faces cannot happen. Its ability to be falsified were such cases to come to light is, of course, part of what

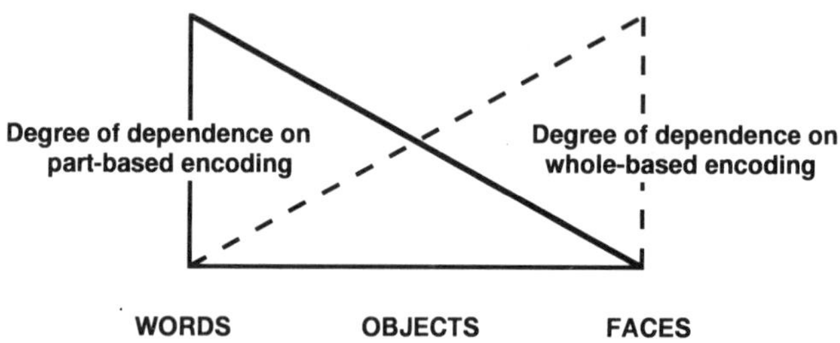

Fig. 4.7 Farah's (1991) hypothesis that the brain uses two structural encoding systems. Part-based encoding is particularly appropriate for words but is also needed for some objects. Whole-based encoding is particularly appropriate for faces but is also needed for some objects.

makes it such a good scientific hypothesis. Unfortunately for Farah's (1991) hypothesis, at least one such case has recently been described, and investigated in detail (Rumiati *et al.* 1994). We must therefore be cautious about its long-term survival. The less parsimonious account that the brain creates separate representations of objects, words, and faces may well prove to be correct.

There are other phenomena which show that the whole-versus part-based encoding hypothesis cannot be more than a part of the overall picture. Again, an example can be taken from studies of face perception, where it has been known for some time that brain injury may compromise different aspects of face processing. For example, dissociations have been described between familiar face recognition and discrimination of unfamiliar faces (Benton 1980). One such report was given by Malone *et al.* (1982), who described two people with different patterns of recovery from brain injury. The first person was initially unable to recognize familiar faces, but had regained this ability by the time that formal neuropsychological tests were given. Yet he was still impaired on tests requiring the matching of views of unfamiliar faces. The second person showed the opposite pattern, with ability to match unfamiliar faces recovering whilst an impairment affecting the recognition of familiar faces persisted. A further type of dissociation concerns recognition of identity and emotional expression; some people who have lost the ability to recognize familiar faces remain able to understand emotional facial expressions (Bruyer *et al.* 1983; Hécaen and Angelergues 1962), and vice versa (Kurucz and Feldmar 1979; Kurucz *et al.* 1979).

Observations of dissociable impairments affecting different aspects of face processing suggest that the brain analyses some of the different social signals we can get from the face independently from each other. In particular, the literature suggests that recognition of emotion, recognition of familiar identity, and the kind of detailed analysis of appearance (age, sex, physical features) needed to match pictures of unfamiliar faces are carried out independently.

The importance of this claim lies in the fact that these different social signals (identity, expression, and so on) can be read from a common physical stimulus (the face). As far as we know, some degree of whole-based encoding is needed both for recognition of a face's identity and for recognition of its expression, yet deficits affecting these abilities can dissociate from each other. This implies that Farah's (1991) emphasis on physical properties of the visual stimulus is incomplete, and we need to give at least as much importance to the social purposes for which different types of information are used.

Recognition of a familiar identity is important because it allows the retrieval of previously stored knowledge about that person, allowing us to moderate our interactions in terms of what we already know from past experience. Analysis of appearance is probably more important when meeting someone for the first time, since it allows reasonably valid inferences about social characteristics such as age and sex, and it also has a more pernicious but subjectively unavoidable influence on attributions concerning intelligence, personality, and so on (Lewicki 1986). In contrast, facial expressions need to be monitored from moment to moment, to detect subtle or not so subtle signs of another person's interest, and their intentions and emotions.

However, the evidence pointing toward dissociable deficits affecting different aspects of face processing has mainly come from isolated reports, often based on quite different tests, and my colleagues and I felt that a more systematic investigation was desirable. To this end, we tested a group of 34 ex-servicemen who had sustained unilateral brain injuries during World War II or subsequent military service, using two different tests of familiar face recognition, two tests of unfamiliar face matching, and two tests of facial expression processing (Young *et al.* 1993). Our aim was to find cases of selective impairment, defined as impaired performance on both tests of a given ability and normal performance on all tests of the other abilities.

When considered in terms of accuracy of performance of the different tasks, seven ex-servicemen were found to meet this stringent criterion for selective impairment. Data for three cases (PG, SJ, and GM) are summarized in Fig. 4.8, with one person corresponding to each of the three possible patterns of accuracy. These are again shown as z-scores to indicate the degree of difference from control performance for each task. For accuracy scores, PG was poor at both familiar face recognition tasks, SJ for both unfamiliar face matching tasks, and GM for both facial expression processing tasks.

These data provide evidence consistent with the view that the brain computes different social signals from the face independently from each other. However, it is desirable also to show selective impairment not only for performance accuracy, but also to show that the relatively preserved abilities are achieved in the 'normal' manner, rather than through ingenious compensatory strategies. This point has been highlighted by Newcombe (1979), who noticed that one of her patients performed face matching tasks quite accurately, but he took a long time and seemed to rely on rather unusual strategies. The importance of these observations was confirmed by asking him to perform a matching task in which the hairlines were masked by an oval frame, to eliminate his preferred hair-matching strategy; his performance accuracy then deteriorated dramatically.

The desirable ideal is thus to find not only normal levels of performance accuracy, but also normal speed, since compensatory strategies usually need extra time. Since Young *et al.* (1993) had data for response times on four of the six tasks they used, they looked at these alongside performance accuracy. When this was done, a rather different pattern emerged (see Fig. 4.8). PG no longer showed a selective impairment; he turned out to be generally slower than controls for all tasks. SJ could carry out expression processing tasks quickly and accurately, but he was slow at identity tasks whether the faces were familiar or unfamiliar. The pattern for GM stood up better, with essentially similar findings of impaired expression processing for both accuracy and speed.

The upshot is that when both accuracy and speed are considered together, we still have clear evidence of dissociable impairments affecting the processing of expression and identity from the face (SJ versus GM), but more equivocal evidence of dissociations between familiar and unfamiliar identities. A caveat remains that stronger evidence of the latter dissociation may be found in future; also, although a combination of normal accuracy and normal speed may represent a desirable ideal, there are many reasons why this might not always be achieved in practice. In other words, whilst a combination of normal accuracy and normal speed can provide strong evidence that a function is well-preserved after brain injury, and unusual strategies will often lead to long response times, it is not necessarily the case that long response latencies must indicate defective performance of the ability in question. For example, they might sometimes simply indicate cautiousness created indirectly by other problems arising from the brain injury.

One way to think about SJ, then, is that he showed intact processing of facial expression, but impaired processing of facial *identity*, whether this was for unfamiliar faces or familiar faces. In contrast, GM showed the opposite pattern. PG showed evidence of a more general impairment when his response times were considered, suggestive of a problem in encoding the face. Interestingly, SJ and PG had right cerebral hemisphere injuries, whereas GMs wound had affected the left hemisphere. It is curious that the person with a selective impairment affecting the processing of facial expressions should have a left hemisphere lesion, since a link between facial expression processing and the right hemisphere has often been noted in other clinical reports (Etcoff 1986). However, the finding seems

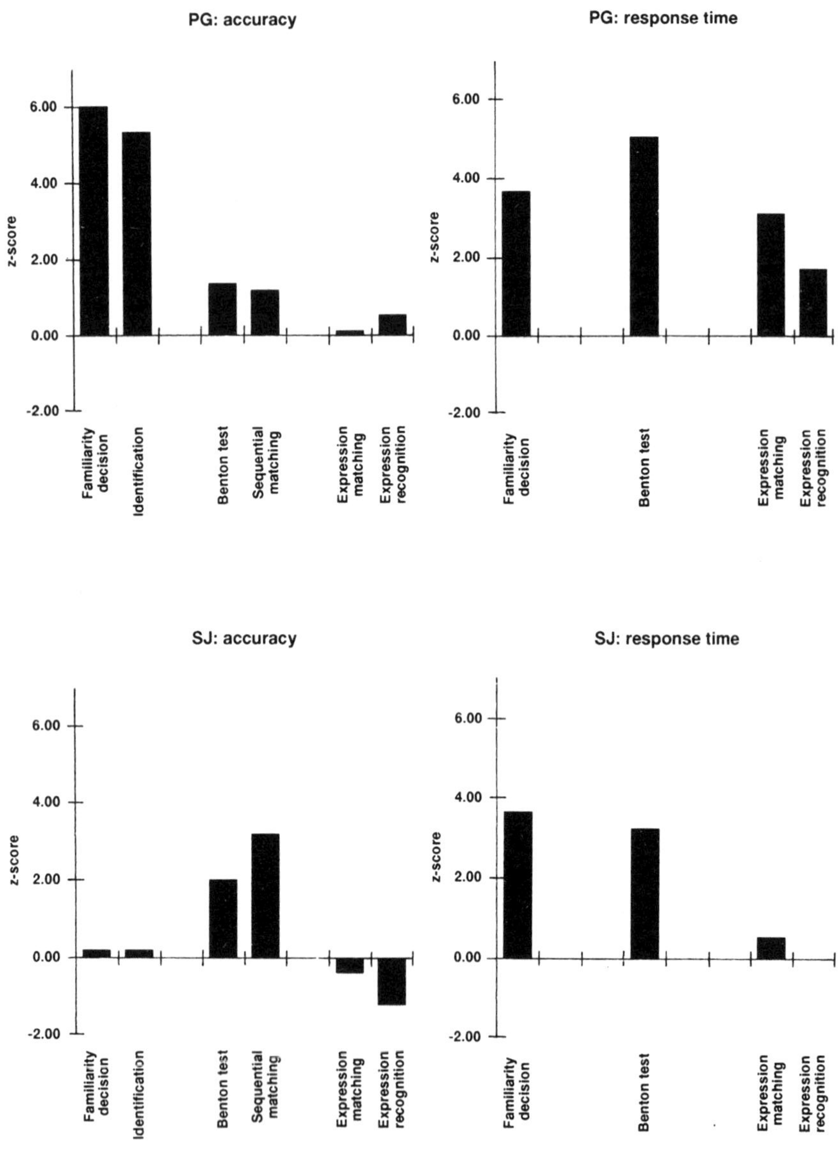
PG: accuracy
PG: response time
SJ: accuracy
SJ: response time
z-score
6.00
4.00
2.00
0.00
-2.00
Familiarity decision
Identification
Benton test
Sequential matching
Expression matching
Expression recognition

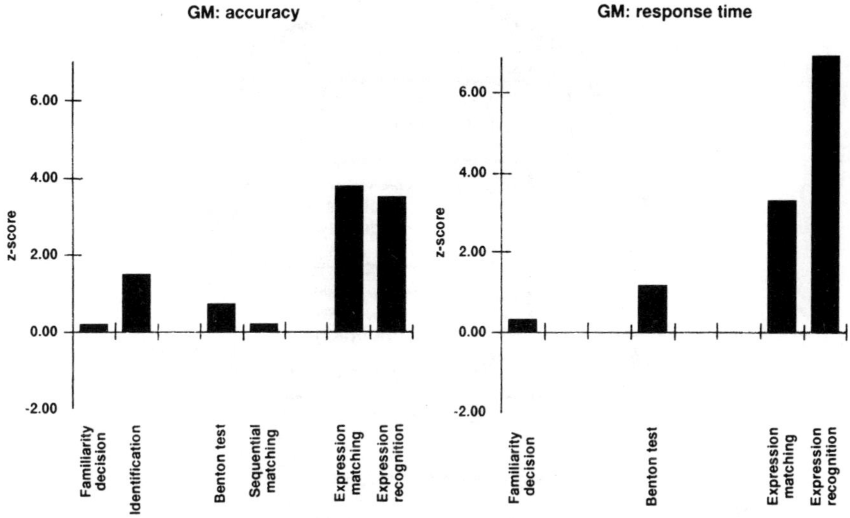

Fig. 4.8 Dissociable impairments of face processing for three ex-servicemen (PG, SJ, GM) studied by Young *et al.* (1993). Performance is expressed as *z*-scores with respect to a neurologically normal control group. There were two tests of familiar face recognition ('familiarity decision' involved deciding whether or not a face is a famous person, 'identification' involved giving appropriate semantic information or names to familiar faces), two tests of unfamiliar face matching (the Benton test, and a sequential matching test in which photographs of unfamiliar faces were presented one after the other), and two tests of facial expression processing ('expression matching' involved finding photographs of different people expressing the same emotion, 'expression recognition' involved deciding which of six emotion labels was most appropriate for a seen face). Accuracy of each person's performance is shown on the left chart, and response times for four of the tests (familiarity decision, Benton test, expression matching, expression recognition) are shown on the right.

to be genuine, because four other cases similar to GM were noted in Young *et. al.*'s (1993) study, and these all had left hemisphere lesions too.

A simple schematic representation of these findings is shown in Fig. 4.9. This indicates abilities identified as susceptible to differential impairment and their relation to each other, and also gives an indication of the extent to which they may depend on the left or the right cerebral hemisphere (in terms of relative positioning along the left–right axis).

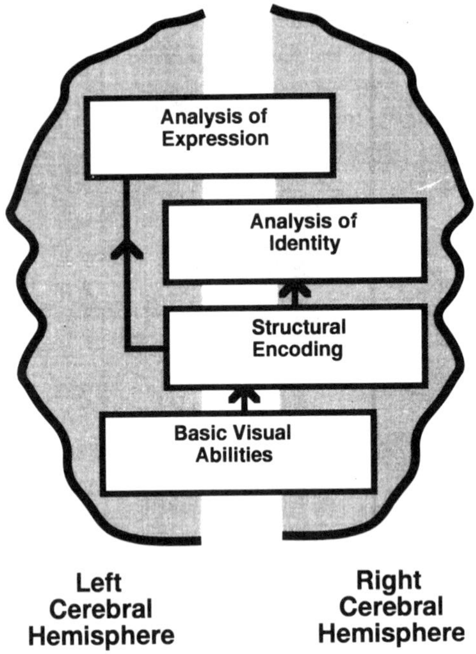

Fig. 4.9 Schematic model of face processing. This shows functional components identified as susceptible to differential impairment and their relation to each other, with an indication of the extent to which they may depend on the left or right cerebral hemisphere (in terms of relative positioning along the left–right axis).

The separation of basic visual abilities from structural encoding in Fig. 4.9 is based on Newcombe *et al.*'s (1989) demonstration of dissociations between impairments affecting the spatial contrast sensitivity function (basic visual abilities) and the face perception task in which people had to decide whether or not a stimulus was a properly organized face (structural encoding). Since Newcombe *et al.* (1989) found that equivalent impairments of the spatial contrast sensitivity function could follow unilateral brain injury affecting either of the cerebral hemispheres, basic visual abilities are represented as bilaterally organized. In contrast, Newcombe *et al.* (1989) only found impairments of face perception among their ex-servicemen with right hemisphere lesions. Hence structural encoding is represented as more dependent on the right than the left hemisphere in Fig. 4.9.

The principal finding of Young *et. al.*'s (1993) study was the double dissociation between selective impairments affecting the analysis of identity and the analysis of expression from the face. So analysis of expression and analysis of identity are represented as independent functions in Fig. 4.9, since they can be fully dissociable from each other. Analysis of expression is represented as more dependent on the left hemisphere because the cases of selective impairment on expression tasks all involved left hemisphere lesions, and analysis of identity is represented as more dependent on the right hemisphere because cases with impaired analysis of identity had right hemisphere lesions. The finding of right hemisphere involvement in the processing of identity is consistent with several other reports but, as we have noted, the finding of left hemisphere involvement in the analysis of expression is more unusual, and needs to be interpreted with care. It does not necessarily mean that the left hemisphere outperforms the right hemisphere in everything that is involved in facial expression analysis, but it does mean that it makes a distinct contribution to the process. It may thus be possible in future to achieve further fractionation of impairments affecting the analysis of facial expressions into different basic forms, and relate these to a more detailed model of expression processing.

This point highlights a potential weakness of multiple-case studies, which is that the investigation of the exact reasons for each person's problems is necessarily somewhat limited. Although this limitation is offset by important strengths of the multiple-case approach, a considerable amount of further information about recognition of identity and recognition of emotion from the face has come from more detailed studies of individual cases.

RECOGNITION OF IDENTITY

The most extensively investigated impairment of face recognition is prosopagnosia, in which a brain-injured person who is still able to see loses the ability to recognize the faces of familiar people that she or he can recognize from non-facial cues, such as voice or name. Notice that this rules out as possible causes of the face recognition impairment both blindness (the person can still see, and in many published cases can see quite well) and general intellectual impairment (people are still recognized from non-facial cues). Yet even the most familiar faces may go unrecognized; famous people, friends, family, and the person's own face when seen in a mirror (Hécaen and Angelergues 1962; Meadows 1974).

The impairment of familiar face recognition found in prosopagnosia is more severe than the impairments documented for the ex-servicemen shown in Fig. 4.8. Even so, people with prosopagnosia know when they are looking at a face, and can often describe its features. Some can interpret facial expressions accurately. Some can match views of unfamiliar faces accurately, though they may often rely on idiosyncratic strategies. The ability to recognize visual objects other than faces may remain quite good in some cases, and the spatial contrast sensitivity function can be normal. Many prosopagnosic patients are able to read without apparent difficulty.

Prosopagnosia is associated with occipito-temporal lesions; these are often bilateral, but cases in which there seems to be only a unilateral lesion affecting the right hemisphere have been reported (De Renzi *et al.* 1994).

An issue which has been widely discussed concerns the relation of prosopagnosia to other impairments of visual recognition, and whether or not the defect is specific to faces. We have seen that prosopagnosia involves a very severe impairment of face recognition, yet a number of people with prosopagnosia can read and can recognize many everyday objects, so the deficit does not compromise all aspects of visual recognition. But is it only face recognition that is impaired?

In most cases, the answer is clearly 'no'. For example, recognition of familiar buildings, flowers, types of car, or animal species may all pose difficulties. Like faces, these are all visual categories with a high degree of inter-item similarity. For instance Bornstein (1963) described a prosopagnosic person who had also lost the ability to identify species of birds which had formerly been well-known to her. She commented that 'all the birds look the same'. Similarly, Bornstein *et al.* (1969) reported the case of a prosopagnosic farmer who could no longer identify his own livestock. Previously he had been able to recognize his cows as individuals, but he could no longer do this.

There seem to be at least two factors involved in such reports. First, these people had some special expertise with another visual category. When they became prosopagnosic, they lost this expertise as well as their expertise with faces. Second, what they seem to have lost is the ability to recognize the individual members of categories that contain several items of similar appearance. Ability to identify the general category to which the items belong (faces, birds, cows, and so on) is preserved, but ability to achieve within-category recognition is lost.

Although the hypothesis that what is lost in prosopagnosia is the ability to recognize the individual members of categories that contain several items of similar appearance is an attractive one, it no longer fits all the evidence. Instead of being consistently present, deficits for visual categories other than faces tend to vary from case to case. Moreover, there are also a few reports of highly circumscribed impairments. One of De Renzi's (1986) cases could find his own belongings when they were mixed in with similar objects, identify his own handwriting, pick out a Siamese cat among photographs of other cats, recognize his own car without reading the number plate, and sort domestic coins from foreign coins. In all of these tasks he was therefore able to identify individual members of visual categories with high inter-item similarity, yet he could not recognize the faces of relatives and close friends. Just as strikingly, the deficit can affect mainly *human* faces; when McNeil and Warrington's (1993) prosopagnosic patient took up farming, he was able to learn to recognize his sheep, and correctly identified several of them from photographs of their faces!

Although they are exceptionally rare, such cases imply that the possibility of face-specific deficits must be taken seriously. We therefore need to ask why this can happen. The most common type of explanation emphasizes some aspect of the visual demands of face recognition; faces form a category with a particularly large number of very similar exemplars (Gross 1992), they place special demands on whole-based encoding mechanisms (Farah 1991), and so on. However, I think that it will prove to be the semantic properties of faces which are the principal cause. It is the need to access previously stored knowledge of familiar individuals rapidly that may dictate the creation of a domain-specific recognition mechanism.

The force of this argument can be seen by considering neuropsychological cases which involve loss of memory for people. Ellis *et al.* (1989) worked with KS, who suffered severe loss of memory for people following a history of epilepsy and a right anterior temporal lobectomy. This was quite different from a prosopagnosia, since people were poorly recognized from their faces or from their names, suggesting a more central loss of information. Yet KS was highly intelligent and certainly not amnesic (her Memory Quotient of 122 was in line with her IQ of 119). She also showed no generalized semantic memory deficit; for example being able to correctly classify a penguin as a bird even though she could not remember that Tchaikovsky was a composer. However, although her problem was strikingly circumscribed, it was not entirely restricted to people; KS was also poor at recognizing famous animals, famous buildings, and familiar product names.

A similar problem has been documented as part of a progressive impairment associated with right temporal lobe atrophy for case VH

(Evans *et al.* 1995). VH initially presented symptoms of prosopagnosia, being unable to recognize faces of many people whose names she recognized with ease, but on follow-up 9 months later she was found to be impaired at recognizing both faces and names, even though she remained highly intelligent (IQ 117 on initial presentation, 119 at follow-up) and did not show any pronounced deterioration of other cognitive skills.

For a member of a social species, inability to remember things about other people can have a devastating impact. Our observations of KS (Ellis *et al.* 1989) revealed her to be severely handicapped in everyday life. Conversations had to be kept to straightforward concrete matters, since any reference to other people or the things they had done could rapidly lead to incomprehension. KS seldom initiated conversations herself, since without being able to remember much about the person she was talking to, she could think of little to say. She worked in a library, but had to be taken off the front desk and given clerical duties when she was found to be unable to answer inquiries because she no longer recognized the names of famous authors.

RECOGNITION OF EMOTION

Although there are many reports of impaired recognition of familiar faces after brain injury, problems in recognizing facial expressions of emotion have received less attention. There are a number of group studies documenting the existence of such impairments (Etcoff 1986), but few detailed investigations of individual cases. This position is now changing, with increased interest evident in the role of the frontal lobes in emotion (Hornak *et al.* 1996) and in the contribution of the amygdala, a medial temporal lobe structure known to be important in social behaviour and emotion in non-human species (Aggleton 1993; Brothers 1992).

In prosopagnosia, the recognition of all familiar faces is compromised to some degree, given due allowance for the degree of familiarity; brain injury has been found to affect ability to recognize politicians but not pop stars, or only to impair recognition of particular individuals. It thus seems a natural starting point to assume by analogy that impairments of facial expression recognition will compromise all emotions. Considerable interest has therefore been generated by a recent claim that the recognition of fear is more affected than other basic emotions following bilateral amygdala damage (Adolphs *et al.* 1994, 1995).

My colleagues and I have explored this claim with DR, a woman who underwent bilateral amygdala surgery to alleviate otherwise intractable epilepsy. We found that the recognition of identity was relatively preserved; DR could recognize pre-morbidly familiar faces without difficulty, and faces of people encountered since her operations could also be recognized if they were well-known to her (Young *et al.* 1995). Our findings showed that in contrast DR was poor at recognizing facial expressions of emotion, but we did not initially investigate whether some emotions were more affected than others (Young *et al.* 1995, 1996).

As a starting point for investigating specific emotions, DR was shown ten examples of facial expressions of each of the six basic emotions (happiness, sadness, anger, fear, surprise, and disgust) from the Ekman and Friesen (1976) series, and asked to categorize each expression with the appropriate emotion label (Calder *et al.* 1996*b*). Figure 4.10 charts the result, with DRs performance expressed as a percentage of control performance for each emotion. Scores of 100 % indicate that DR recognizes an emotion as well as controls recognize it, lower scores indicate increasing degrees of impairment. DR showed a striking problem in the recognition of fear.

To investigate this further, we tested DRs recognition of photographic quality images interpolated ('morphed') between prototype expressions for these six emotions, to create a hexagonal continuum (running from happiness to surprise to fear to sadness to disgust to anger to happiness). Morphed expressions were generated by blending two pictures of the same person posing different expressions (for example anger and happiness) in physically equal steps, to produce a continuum ranging between the two prototypes. The technique used was initially developed by Benson and Perrett (1991), but has been extended and improved; more detail can be found elsewhere (Calder *et al.* 1996*a*). It involves locating the positions of a large number of specified feature points on each of two prototype images. The shape of one prototype image can then be moved toward the shape of the other prototype, and a continuous-tone (photographic quality) image can be created by 'stretching' both of the prototype images (as if they were printed on a rubber sheet) to this new shape, so that all points representing the same feature are aligned across images, and blending them with the appropriate weight.

Our 'emotion hexagon' is shown in Fig. 4.11. Normal perceivers see this as falling into six distinct emotion sections, with relatively sharp boundaries separating the different regions. In Fig. 4.11, the surprise region corresponds to the two rightmost morphed images

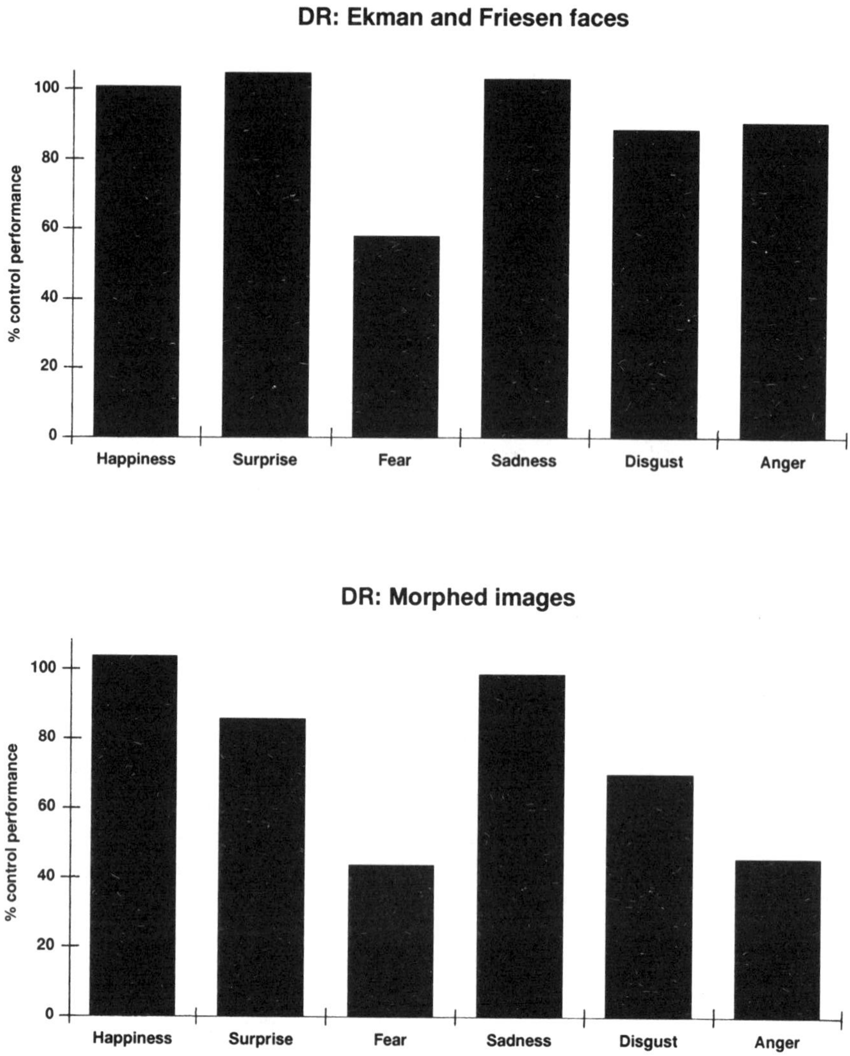

Fig. 4.10 Recognition of basic emotions from facial expressions by DR. Performance for recognizing facial expressions is represented as a percentage of control performance for each emotion. The two tests used faces from the Ekman and Friesen (1976) series (upper chart), and for recognizing morphed face images shown in Fig. 4.11 (lower chart).

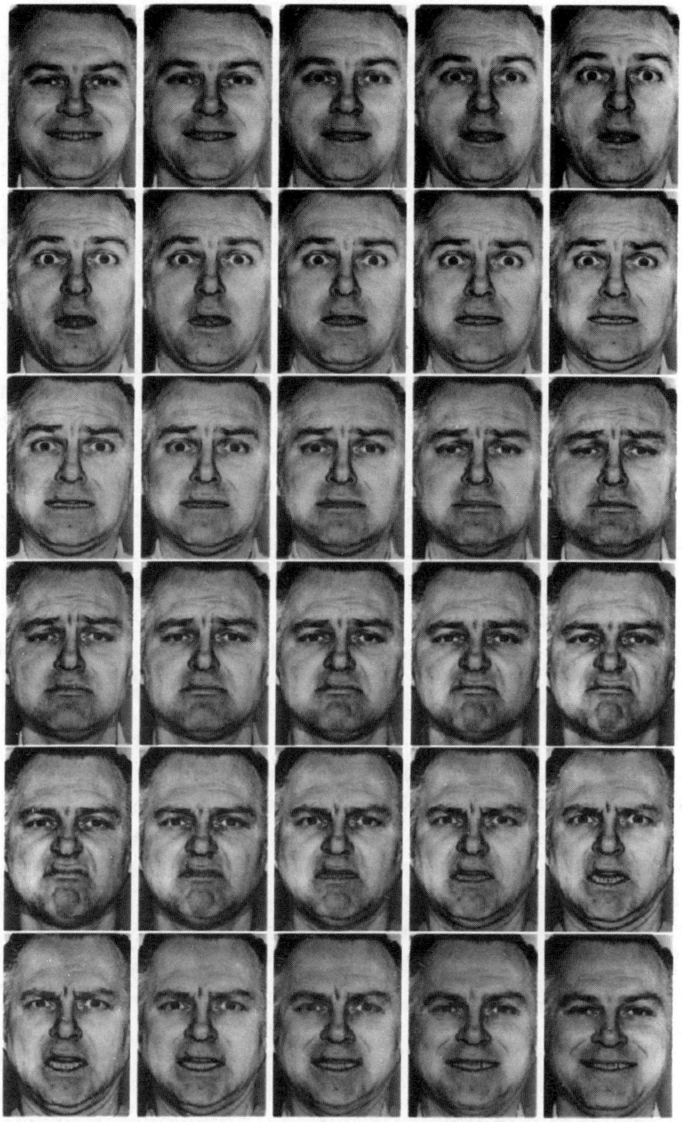

Fig. 4.11 Emotion hexagon. Going from left to right, the columns show morphed images located at 90%, 70%, 50%, 30%, and 10% distances between prototype expressions taken from the Ekman and Friesen (1976) series. These prototypes are not shown in the figure; all of the images are morphs. From top to bottom, the continua shown in each row are happiness – surprise (top row), surprise – fear (second row), fear – sadness (third row), sadness – disgust (fourth row), disgust – anger (fifth row), anger – happiness (bottom row).

in the top row and the two leftmost morphs in the second row, the fear region to the two rightmost morphs in the second row and the two leftmost morphs in the third row, the sadness region to the two rightmost morphs in the third row and the two leftmost morphs in the fourth row, the disgust region to the two rightmost morphs in the fourth row and the two leftmost morphs in the fifth row, the anger region to the two rightmost morphs in the fifth row and the two leftmost morphs in the bottom row, and the happiness region to the two rightmost morphs in the bottom row and the two leftmost morphs in the top row. Other images are seen as intermediate between the two corresponding prototypes.

We presented the 30 images shown in Fig. 4.11 one at a time in random order, in a number of blocks of trials, then calculated the total number of times DR identified all of the four morphs in each region with the appropriate emotion label. These values were compared to the mean number of times control subjects produced the appropriate responses in the same region. As a guide, DRs performance is shown in Fig. 4.10 as a percentage of control performance for each emotion region. She was again found to be poor at recognizing fear, but also anger and (to a lesser extent) disgust. Happiness, sadness, and surprise were relatively unaffected.

Taking together the results summarized in Fig. 4.10, it is clear that DR was more impaired at recognizing some emotions than others, and especially fear. Her impaired recognition of fear and anger implies that the amygdala plays a particularly significant role in the recognition of these expressions. This is consistent with evidence from neurophysiological studies pointing to a specific role for the amygdala in the appraisal of danger and the emotion of fear (LeDoux 1995). In addition, Halgren (1992) has pointed out that electrical stimulation of the amygdala in humans can evoke a variety of phenomena but when an emotional sensation is experienced, it is invariably fear.

We noted that impairments of the recognition of identity found in cases of prosopagnosia affect all familiar faces, so it had seemed natural to assume by analogy that impairments of facial expression recognition will compromise all emotions. This assumption was clearly incorrect. Instead, the important factor may be that facial identities must all be learnt, whereas babies' recognition and production of certain facial expressions seems to be innate (Darwin 1872; Field *et al.* 1982). Hence, the neural substrates of emotion may include templates for specific emotions. Moreover, emotions may have evolved in a piecemeal manner, with differing neural substrates for different emotions. Ekman (1992) has cogently summarized the powerful arguments for a set of basic human emotions, shaped by evolution.

These evolutionary arguments apply most directly to the experience of emotion, whereas what our results for DR demonstrated was impaired recognition of emotion in the faces of others. This is consistent with the idea that for certain basic emotions there is a close mapping of perceptual mechanisms and corresponding emotional experience (Brothers 1989; Niedenthal and Halberstadt 1995). Mechanisms for perceiving emotions in others may be closely linked to those involved in actually experiencing the equivalent emotions oneself because this has evolved as one of the main ways we learn our emotional reactions; many of the things which frighten us do so not because we have direct experience, but because we have learnt from the reactions of others. Fear is a very contagious emotion. This has clear biological advantages, since it allows us to learn about different types of danger without being harmed ourselves.

CONSCIOUS AND NON-CONSCIOUS REPRESENTATIONS

We have seen how neuropsychological evidence can be used to identify dissociable types of perceptual impairment, and how these in turn tell us something about how the brain represents what is going on in our social and physical environments. So far, though, all of our discussion has involved direct tests. If we want to know whether someone who has suffered a brain injury is poor at recognizing familiar faces, we show them some faces and ask who they are. If we want to know if they are poor at recognizing facial expressions, we show them some facial expressions and ask which emotion they correspond to, and so on.

Intuitively, direct tests, in which one asks explicitly about the ability in question, are the obvious way to proceed. But many recent findings have demonstrated that a different picture can emerge if one shifts to indirect tests, in which the ability of interest only arises ostensibly as an almost incidental feature of the task, or in which people are strongly encouraged to guess when they claim that they have no idea what the answer might be (Young 1996).

The most well-known of such examples involves the phenomenon of blindsight (Weiskrantz 1986), in which brain-injured people who deny seeing a visual stimulus presented in a defective part of their field of vision may none the less show accurate appreciation of some of its sensory qualities when given forced-choice tasks and encouraged to guess. Weiskrantz (1986) reports many such examples, including pointing or looking at the locations of subjectively unseen stimuli, and discriminating the orientations of unseen lines.

Such phenomena force us to think hard about the differences between conscious and non-conscious representations (Young and Block 1996). For present purposes, the most important example is covert recognition in prosopagnosia (Young 1994). This was demonstrated by Bauer (1984) who measured skin conductance (GSR) responses whilst a prosopagnosic person (LF) viewed a familiar face and listened to a list of names. When the name belonged to the face being viewed, there was a greater skin conductance change than when it was someone else's name. Yet if LF was asked to choose which name in the list went with the face, his performance was at chance level. The face recognition reflected in the GSR changes was entirely non-conscious.

As well as physiological indices such as GSR, covert recognition can be demonstrated with behavioural measures. For several years, my colleagues and I have worked with PH, who suffered a severe closed head injury in a road accident (de Haan *et al.* 1987*a*). PH can read, and recognize printed names, but he recognizes very few faces overtly. In even the simplest of overt recognition tasks, where he was just asked to pick which of two faces (one unfamiliar, one famous) he had ever seen before, PH performed at chance level, with 65/128 correct choices (Young and de Haan 1988).

Despite this almost complete absence of overt recognition of familiar faces found for direct tests, PH has shown covert recognition effects in many behavioural tasks. These demonstrations have included faster reaction times to familiar than unfamiliar faces for face matching (de Haan *et al.* 1987*a*), priming from unrecognized faces onto the recognition of subsequently presented names (Young *et al.* 1988), and better learning of true than untrue face–name pairings (de Haan *et al.* 1987*a*, 1991; Young and de Haan 1988).

A further example involves face-name interference. Examples of stimuli are shown in Fig. 4.12. The person is asked to classify the names as those of politicians or non-politicians as quickly as possible, ignoring the distractor faces. Trials such as that at the top of Fig. 4.12, where the name of the politician Geoffrey Howe is combined with a distractor face belonging to another politician (Tony Benn), are easier than the lower type of trial where the face (the pop star Paul McCartney) cannot be ignored and interferes with performance because it comes from a different semantic category to the name that has to be classified. PH showed such interference from distractor faces onto name classification even when (unlike the examples in Fig. 4.12) the face categories were carefully matched so that they did not differ for surface features like hair-length or age (de Haan *et al.* 1987*a*, *b*). Under such circumstances, interference must have been based on some form of non-conscious recognition of the identities of the faces.

Such findings can be simulated using a computer model of processes involved in face recognition if the connection strengths of certain links in the model are halved (Burton *et al.* 1991). If the model is of the interactive activation type used by Burton *et al.* (1991), the consequence of such damage is that an input will no longer be able to raise its corresponding nodes above threshold (that is the face is not recognized), yet excitation can still pass at below-threshold levels, leading to the observed patterns of performance.

Whilst this gives us an insight into the type of mechanism which might underlie covert recognition effects, and hence makes the findings less mysterious, it remains the case that they have important general implications. In particular, it is very striking that people like PH do not act on their preserved capacities. Instead, the findings fit the common-sense idea that awareness supports intentional action. For example despite the extensive range of covert effects demonstrated in the laboratory, patients like PH do not act as if they recognize faces in everyday life. This is not a trivial point; one could imagine that PH might find him or herself greeting people in the street without knowing why, yet this does not seem to happen.

What, then, is the functional significance of covert effects? It seems that they reflect aspects of visual information processing that can proceed without conscious experience. Why?

Part of the answer lies in the delicate balance between speed of response and flexibility of response. Consider the favourite textbook comparison to the frog, whose retinal outputs include bug detectors which allow it to make very fast responses to flies. Our own visual system is much less specialized at the retinal level, allowing us increased flexibility of response, but at the cost of loss of speed. Some types of covert effect can thus be seen to reflect automatic components built in to the visual system to mitigate some of this loss of speed; for example by predicting what is likely to follow, and setting up preparatory responses.

The costs of flexibility become especially marked when time is spent not only in constructing an adequate representation, but also in weighing up possible alternative actions. An obvious way

Fig. 4.12 Examples of stimuli used in interference task. The person is asked to classify the names as those of politicians or non-politicians as quickly as possible. When the name is combined with a distractor face belonging to another person from the same category (as in the upper example) it is easier to ignore the distractor face than when the face comes from a different semantic category to the name that has to be classified (lower example).

to balance the competing demands of flexibility and speed is to allow most actions to run off under automatic control, and to involve mechanisms which can allow greater choice only when these are needed. This is a particularly convenient solution for the nervous system, in which many actions (like breathing) can be safely left under 'automatic pilot' much of the time, and only require occasional conscious intervention (if you are about to stick your head underwater, and so on). It is doubly convenient because nervous systems must evolve, and one way to do this is to add extra levels of control to mechanisms that are already sufficient for many purposes.

CONCLUSIONS

The findings reviewed here show the importance of understanding the consequences of brain injury, and how these can have profound implications for our conception of how mental abilities are represented in the brain. We have seen how brain injury can have highly selective effects on our mental abilities, with dissociations between ability to recognize identity or expression from the face, and even more specific defects affecting the recognition of some emotions more than others. Such findings imply that the brain is as much an organ devoted to the perception of the social as the physical environment (Brothers 1992). Other examples which have not been reviewed here reinforce this perspective; dissociable deficits have also been found in the perception of eye gaze direction (Campbell *et al.* 1990), a signal with widespread social and communicative significance, and in lipreading speech signals from movements of the mouth and tongue (Campbell 1986). Failures of social perception in neurodevelopmental disorders such as autism have also been the subject of detailed investigation (Baron-Cohen and Ring 1994).

So a remarkable lesson from neurophysiological and neuropsychological work is that our perception of our physical environment depends on mechanisms which are neurologically somewhat fragmented. After brain injury, perceptual experience can fractionate, with some forms of experience impaired, others intact. This raises interesting questions about the nature of subjective experience, and why it seems so unified to the perceiver. How can the world look the way it does to us if different parts of our brains are responsible for different perceptual qualities? These issues remain unresolved, but the consensus is that the evidence that the brain parcels out the visual input to specialist areas is very strong, whereas there are only weak grounds for thinking there might be other regions devoted to 'putting it all back together'.

Returning to our starting point, we noted that the goal of perception is to create representations of external events in ways that permit effective actions in the world that an organism inhabits. We have seen how this insight may help in accounting for some of the differences between conscious and non-conscious representations. But we need to think not only in terms of representing the physical world, but the social world too – the moods, identities, and intentions of others. This is essential to any social species. Our social abilities involve not just what we learn from birth, but have been shaped by many evolutionary developments. We should perhaps be less surprised than we at present seem to be that the brain is as much devoted to making sense of the interpersonal as the physical environment.

REFERENCES

Adolphs, R., Tranel, D., Damasio, H., and Damasio, A. (1994). Impaired recognition of emotion in facial expressions following bilateral damage to the human amygdala. *Nature*, **372**, 669–72.

Adolphs, R., Tranel, D., Damasio, H., and Damasio, A.R. (1995). Fear and the human amygdala. *Journal of Neuroscience*, **15**, 5879–91.

Aggleton, J.P. (1993). The contribution of the amygdala to normal and abnormal emotional states. *Trends in Neurosciences*, **16**, 328–33.

Baron-Cohen, S. and Ring, H. (1994). A model of the mindreading system: neuropsychological and neurobiological perspectives. In *Children's early understanding of mind: origins and development* (ed. C. Lewis and P. Mitchell), pp. 183–207. Lawrence Erlbaum, Hove, East Sussex.

Bauer, R.M. (1984). Autonomic recognition of names and faces in prosopagnosia: a neuropsychological application of the guilty knowledge test. *Neuropsychologia*, **22**, 457–69.

Benson, P.J. and Perrett, D.I. (1991). Synthesising continuous-tone caricatures. *Image and Vision Computing*, **9**, 123–9.

Benton, A.L. (1980). The neuropsychology of facial recognition. *American Psychologist*, **35**, 176–86.

Bornstein, B. (1963). Prosopagnosia. In *Problems of dynamic neurology* (ed. L. Halpern), pp. 283–318. Hadassah Medical School, Jerusalem.

Bornstein, B., Sroka, H., and Munitz, H. (1969). Prosopagnosia with animal face agnosia. *Cortex*, **5**, 164–9.

Brothers, L. (1989). A biological perspective on empathy. *American Journal of Psychiatry*, **146**, 10–19.

Brothers, L. (1992). Perception of social acts in primates: cognition and neurobiology. *Seminars in the Neurosciences*, **4**, 409–14.

Bruyer, R., Laterre, C., Seron, X., Feyereisen, P., Strypstein, E., Pierrard, E., *et al.* (1983). A case of prosopagnosia with some preserved covert remembrance of familiar faces. *Brain and Cognition*, **2**, 257–84.

Burton, A.M., Young, A.W., Bruce, V., Johnston, R., and Ellis, A.W. (1991). Understanding covert recognition. *Cognition*, **39**, 129–66.

Calder, A.J., Young, A.W., Perrett, D.I., Etcoff, N.L., and Rowland, D. (1996*a*). Categorical perception of morphed facial expressions. *Visual Cognition*, **3**, 101–37.

Calder, A.J., Young, A.W., Rowland, D., Perrett, D.I., Hodges, J.R., and Etcoff, N.L. (1996*b*). Facial emotion recognition after bilateral amygdala damage: differentially severe impairment of fear. *Cognitive Neuropsychology*, **13**, 699–745.

Campbell, R., Landis, T., and Regard, M. (1986). Face recognition and lipreading: a neurological dissociation. *Brain*, **109**, 509–21.

Campbell, R., Heywood, C.A., Cowey, A., Regard, M., and Landis, T. (1990). Sensitivity to eye gaze in prosopagnosic patients and monkeys with superior temporal sulcus ablation. *Neuropsychologia*, 1123–42.

Cowey, A. (1994). Cortical visual areas and the neurobiology of higher visual processes. In *The neuropsychology of high-level vision: collected tutorial essays* (ed. M.J. Farah and G. Ratcliff), pp. 3–31. Lawrence Erlbaum, Hillsdale, New Jersey.

Darwin, C. (1872). *The expression of the emotions in man and animals*. John Murray, London.

de Haan, E.H.F., Young, A., and Newcombe, F. (1987*a*). Face recognition without awareness. *Cognitive Neuropsychology*, **4**, 385–415.

de Haan, E.H.F., Young, A., and Newcombe, F. (1987*b*). Faces interfere with name classification in a prosopagnosic patient. *Cortex*, **23**, 309–16.

de Haan, E.H.F., Young, A.W., and Newcombe, F. (1991). Covert and overt recognition in prosopagnosia. *Brain*, **114**, 2575–91.

De Renzi, E. (1986). Current issues in prosopagnosia. In *Aspects of face processing* (ed. H.D. Ellis, M.A. Jeeves, F. Newcombe, and A. Young), pp. 243–52. Martinus Nijhoff, Dordrecht.

De Renzi, E., Perani, D., Carlesimo, G.A., Silveri, M.C., and Fazio, F. (1994). Prosopagnosia can be associated with damage confined to the right hemisphere – an MRI and PET study and a review of the literature. *Neuropsychologia*, **32**, 893–902.

Ekman, P. (1992). An argument for basic emotions. *Cognition and Emotion*, **6**, 169–200.

Ekman, P. and Friesen, W.V. (1976). *Pictures of facial affect*. Consulting Psychologists Press, Palo Alto, California.

Ellis, A.W., Young, A.W., and Critchley, E.M.R. (1989). Loss of memory for people following temporal lobe damage. *Brain*, **112**, 1469–83.

Ellis, H.D. (1983). The role of the right hemisphere in face perception. In *Functions of the right cerebral hemisphere* (ed. A.W. Young), pp. 33–64. Academic Press, London.

Etcoff, N.L. (1986). The neuropsychology of emotional expression. In *Advances in clinical neuropsychology* (ed. G. Goldstein and R.E. Tarter), pp. 127–79. Plenum, New York.

Evans, J.J., Heggs, A.J., Antoun, N., and Hodges, J.R. (1995). Progressive prosopagnosia associated with selective right temporal lobe atrophy: a new syndrome? *Brain*, **118**, 1–13.

Farah, M.J. (1991). Patterns of co-occurrence among the associative agnosias: implications for visual object representation. *Cognitive Neuropsychology*, **8**, 1–19.

Field, T.M., Woodson, R., Greenberg, R., and Cohen, D. (1982). Discrimination and imitation of facial expressions by neonates. *Science*, **218**, 179–81.

Gross, C.G. (1992). Representation of visual stimuli in inferior temporal cortex. *Philosophical Transactions of the Royal Society, London*, **B335**, 3–10.

Halgren, E. (1992). Emotional neurophysiology of the amygdala within the context of human cognition. In *The amygdala* (ed. J.P. Aggleton), pp. 191–228. Wiley-Liss, New York.

Hécaen, H. and Angelergues, R. (1962). Agnosia for faces (prosopagnosia). *Archives of Neurology*, **7**, 92–100.

Hornak, J., Rolls, E.T., and Wade, D. (1996). Face and voice expression identification in patients with emotional and behavioural changes following ventral frontal lobe damage. *Neuropsychologia*, **34**, 247–61.

Kurucz, J. and Feldmar, G. (1979). Prosopo-affective agnosia as a symptom of cerebral organic disease. *Journal of the American Geriatrics Society*, **27**, 225–30.

Kurucz, J., Feldmar, G., and Werner, W. (1979). Prosopo-affective agnosia associated with chronic organic brain syndrome. *Journal of the American Geriatrics Society*, **27**, 91–5.

LeDoux, J.E. (1995). Emotion: clues from the brain. *Annual Review of Psychology*, **46**, 209–35.

Lewicki, P. (1986). Processing information about covariations that cannot be articulated. *Journal of Experimental Psychology: Learning, Memory and Cognition*, **12**, 135–46.

Malone, D.R., Morris, H.H., Kay, M.C., and Levin, H.S. (1982). Prosopagnosia: a double dissociation between the recognition of familiar and unfamiliar faces. *Journal of Neurology, Neurosurgery, and Psychiatry*, **45**, 820–2.

McNeil, J.E. and Warrington, E.K. (1993). Prosopagnosia: a face specific disorder. *Quarterly Journal of Experimental Psychology*, **46A**, 1–10.

Meadows, J.C. (1974). The anatomical basis of prosopagnosia. *Journal of Neurology, Neurosurgery, and Psychiatry*, **37**, 489–501.

Newcombe, F. (1979). The processing of visual information in prosopagnosia and acquired dyslexia: functional versus physiological interpretation. In *Research in psychology and medicine, Vol 1* (ed. D.J. Oborne, M.M. Gruneberg, and J.R. Eiser), pp. 315–22. Academic Press, London.

Newcombe, F. and Russell, W.R. (1969). Dissociated visual perceptual and spatial deficits in focal lesions of the right hemisphere. *Journal of Neurology, Neurosurgery, and Psychiatry*, **32**, 73–81.

Newcombe, F., Ratcliff, G., and Damasio, H. (1987). Dissociable visual and spatial impairments following right posterior cerebral lesions:

clinical, neuropsychological and anatomical evidence. *Neuropsychologia*, **25**, 149–61.

Newcombe, F., de Haan, E.H.F., Ross, J., and Young, A.W. (1989). Face processing, laterality, and contrast sensitivity. *Neuropsychologia*, **27**, 523–38.

Niedenthal, P.M. and Halberstadt, J.B. (1995). The acquisition and structure of emotional response categories. In *The psychology of learning and motivation: advances in research and theory Vol 33*. (ed. D.L. Medin), pp. 23–64. Academic Press, New York.

Rumiati, R.I., Humphreys, G.W., Riddoch, M.J., and Bateman, A. (1994). Visual object agnosia without prosopagnosia or alexia: evidence for hierarchical theories of visual recognition. *Visual Cognition*, **1**, 181–225.

Sergent, J. (1983). Role of the input in visual hemispheric asymmetries. *Psychological Bulletin*, **93**, 481–512.

Sergent, J. (1984). Inferences from unilateral brain damage about normal hemispheric functions in visual pattern recognition. *Psychological Bulletin*, **96**, 99–115.

Ungerleider, L.G. and Mishkin, M. (1982). Two cortical visual systems. In *Analysis of visual behavior* (ed. D.J. Ingle, M.A. Goodale, and R.J.W. Mansfield), pp.549–86. MIT Press, Cambridge, Massachusetts.

Weiskrantz, L. (1986). *Blindsight: a case study and implications*. Oxford University Press, Oxford.

Young, A.W. (1987). Cerebral hemisphere differences and reading. In *Cognitive approaches to reading* (ed. J. Beech and A. Colley), pp.139–68. Wiley, Chichester.

Young, A.W. (1994). Covert recognition. In *The neuropsychology of high-level vision: collected tutorial essays* (ed. M.J. Farah and G. Ratcliff), pp.331–58. Lawrence Erlbaum, Hillsdale, New Jersey.

Young, A.W. (1996). Dissociable aspects of consciousness. In *The science of consciousness: psychological, neuropsychological and clinical reviews* (ed. M. Velmans), pp.118–39. Routledge, London.

Young, A.W. and Block, N. (1996). Consciousness. In *Unsolved mysteries of the mind: tutorial essays in cognition* (ed. V. Bruce), pp. 149–79. Erlbaum (UK) Taylor and Francis, Hove, East Sussex.

Young, A.W. and de Haan, E.H.F. (1988). Boundaries of covert recognition in prosopagnosia. *Cognitive Neuropsychology*, **5**, 317–36.

Young, A.W., Hay, D.C., and McWeeny, K.H. (1985*a*). Right cerebral hemisphere superiority for constructing facial representations. *Neuropsychologia*, **23**, 195–202.

Young, A.W., Hay, D.C., McWeeny, K.H., Ellis, A.W., and Barry, C. (1985*b*). Familiarity decisions for faces presented to the left and right cerebral hemispheres. *Brain and Cognition*, **4**, 439–50.

Young, A.W., Hellawell, D., and de Haan, E.H.F. (1988). Cross-domain semantic priming in normal subjects and a prosopagnosic patient. *Quarterly Journal of Experimental Psychology*, **40A**, 561–80.

Young, A.W., Newcombe, F., de Haan, E.H.F., Small, M., and Hay, D.C. (1993). Face perception after brain injury: selective impairments affecting identity and expression. *Brain*, **116**, 941–59.

Young, A.W., Aggleton, J.P., Hellawell, D.J., Johnson, M., Broks, P., and Hanley, J.R. (1995). Face processing impairments after amygdalotomy. *Brain*, **118**, 15–24.

Young, A.W., Hellawell, D.J., van de Wal, C., and Johnson, M. (1996). Facial expression processing after amygdalotomy. *Neuropsychologia*, **34**, 31–9.

5

Toxins in the environment and human brain disease

David Marsden

Professor David Marsden, DSc, FRCP, FRS, is Professor of Clinical Neurology and Director of the MRC Human Movement and Balance Unit at the National Hospitals for Neurology and Neurosurgery, Queen Square in London, where he began his career in 1968 as Senior House Physician after qualifying at St Thomas's Hospital. In 1972 he was appointed Professor of Neurology at King's College Hospital, a post which he held for fifteen years. He was elected a Fellow of the Royal Society in 1983 and has served on its Council. He is the author of numerous publications on human movement disorders and motor physiology, fields in which he is among the world's leading experts.

The brain is the basis of the mind. The ability to walk on two legs, freeing the hands for the manipulation of tools, to articulate so as to produce speech, which allows structured communication, and to introspect and so communicate our thoughts are remarkable human attributes. All depend upon the exquisite function of the human brain, which we can see operating in life.

The technique of positron emission tomography (PET) has allowed brain scientists to view the thinking and moving brain at work. A radioactive tracer (usually radioactive oxygen) is injected intravenously, and computerized images of radioactivity in the brain are collected while the subject executes a particular task. Comparison of brain radioactivity is made during the task in question, and compared to activity while the subject is quietly resting, or while executing a control task so as to subtract brain action during a specific mental or motor activity. Measurements of changes in blood flow are made and from the increased blood flow those parts of the brain engaged in a specific task, therefore requiring more energy, are identified. For example, when a person executes a simple movement, such as flexing or extending a finger, in response to a metronome signal, the primary motor areas of the brain are activated. In contrast, when the person has to make the same movement but in response to their own initiative, additional parts of the brain are activated, in particular premotor areas in the frontal lobes. When the person is asked to think about this movement but not actually execute it, more anterior parts of the brain are activated, particularly prefrontal regions. Thus one can see the brain thinking of moving, conceiving a movement in response to internal thought, or merely executing the movement.

This powerful technique of PET study of the working human brain has begun to unravel the complexities of organization that underlie our most human characteristics. However, the brain is vulnerable to a large range of environmental insults, which are the topic of this essay. I have chosen a number of different areas to illustrate the theme.

ALZHEIMER'S DISEASE

If the brain deteriorates, 'the human being disappears'. As a clinical neurologist, I see this every week in the tragedy of Alzheimer's disease. Alzheimer's disease is the plague of old age in our community today. The brain suffers from progressive degeneration of its constituent nerve cells. Pathologically, the disease is identified by the increasing appearance of senile plaques and neurofibrillary tangles (Khachaturian 1985). The plaques consist of dystrophic nerve terminals clustered around a core of β-amyloid protein. The neurofibrillary tangles are derived from the microtubular-associated protein tau. Senile plaques and neurofibrillary tangles occur to a mild extent in normal old people, but in those who develop the symptoms of Alzheimer's disease the number increases disproportionately in the cerebral cortex, causing progressive destruction of the thinking brain.

Alzheimer's disease is rare in young people, but exponentially increases in frequency over the age of 50 years. Approximately 1% of those aged between 50 and 70 years develop Alzheimer's disease, but then there is a dramatic rise to about 50% in very elderly people (Jorm 1990; Skoog *et al.* 1993).

Characteristically, the disease starts with impairment of memory. Breakdown of language, with both word finding and comprehension difficulties, and distorted perception of the visuo-spatial world soon emerge (Rossor 1995). To the individual patient, Alzheimer's disease begins as a terrifying experience. Memory fails and social life becomes increasingly impossible. However, sooner or later insight is lost and the patient is relieved of the burden of fear of their impending dissolution. To their family, the illness is a disaster. They are forced to witness the increasing disintegration of the personality of their loved one, with disappearance of the nuances that make up the individual. There can be no greater tragedy than to experience such collapse of those characteristic and lovable features that identify our closest partners, parents, and friends. All this is due to the destruction by the disease process of the human brain. Most patients die within 7 years.

With an aging population throughout the world the number of patients with dementia due to Alzheimer's disease is rising. Based on demographic trends of people aged over 80 years in the United States it is anticipated that there may be 10–15 million people affected by Alzheimer's disease in that country by the middle of the next century (Office of Technology Assessment, US Congress 1987). Global figures suggest that the number may exceed one hundred million worldwide (Whitehouse 1993). In the United Kingdom, there are currently over half a million people predicted to have Alzheimer's disease. Although the disease is uncommon in younger adults, there are believed to be 18 000 individuals in the United Kingdom below the age of 65 years with the condition (Rossor 1995).

A recent estimate of the costs of Alzheimer's disease to the community in England has suggested that a total of £1144 million was spent in 1992, illustrating the burden of this illness to our country (Grey and Fenn 1993). This analysis did not include any indirect costs to family carers over and above the benefits they received from Government agencies. The burden of informal care from family members is substantial.

The cause of Alzheimer's disease is not known. About 1 in 10 cases appear to inherit the disease as the result of a genetic defect. However, most cases are isolated and the question arises as to whether such sporadic cases of Alzheimer's disease may be due to some environmental agent. So far, no convincing environmental toxin has been identified as a cause of Alzheimer's disease. Aluminium has been suggested, for tangles in the brain contain this metal, and aluminium toxicity in renal dialysis units has caused dementia in those with kidney failure, albeit with a different pathology. However, the epidemiological evidence incriminating aluminium as a cause of Alzheimer's disease is far from convincing.

The matter is more complicated, for it may be that an inherited trait, such as apolipoprotein E4, may predispose individuals to the development of the disease. Thus there may be a 'double hit' scenario;

an inherited predisposition combined with an environmental agent, may result in the disease.

This introduces the concept that some brain diseases may be due to our genes, some to a toxic agent in the environment, and some to a combination of the two. Environmental events that cause brain injury vary from the obvious, such as head injury, through infections such as AIDS, to more complex situations, such as Parkinson's disease. The various examples that follow illustrate the point.

HEAD INJURY, THE MOTOR CAR, AND ALCOHOL

About one million individuals in the United Kingdom suffer a head injury per year (Strang *et al.* 1987; Jennett and MacMillan 1981). Almost half of these are children under the age of 15 years, and boys outnumber girls by more than 2 to 1. Most injuries (41%) are due to a fall, and many (20%) are due to an assault. Road traffic accidents are a major source of head injury in any community. Although they are responsible for only 13% of those attending hospital, they account for a third of patients requiring transfer to neurosurgical units and 58% of deaths. The death rate from head injury in the United Kingdom is about 9 per 100 000 per year. Head injury accounts for 1% of all deaths; it is responsible for 15–20% of deaths in young people aged between 5–35 years (Teasdale 1995).

Preventive measures, such as speed limits, legal enforcement of alcohol levels, seat-belts, air bags, and helmets for motor cyclists, have had a major impact on the mortality and morbidity of head injury. But nothing is simple. Motor cyclists who wear helmets now die less frequently of head injury, but the risks of damage to the neck, and consequential paralysis, have increased.

Accident prevention requires modification of behaviour by the public and is effective usually only when enforced by legislation. However, the balance of legal restraint versus individual freedom may be difficult to achieve, and sometimes the best action may not be clear. A recent example is the attempt to encourage a shift in emphasis from the use of petrol to diesel fuel. Toxic emissions from petrol combustion were judged to be deleterious to the environment, particularly carbon monoxide and lead, so catalytic convertors have been fitted to cars, and diesel prices kept low to encourage a shift to cars using the latter. Now there is the increasing environmental hazard of particulates in the air generated more by diesel emissions.

Whether lead in the environment was really the major hazard it was claimed to be is questionable. Lead certainly can damage the brain, as was evident in those children who indulged in licking or eating old lead soldiers or lead-based paints. But low levels of lead exposure may not have been as deleterious as was thought.

Alcohol is, of course a major environmental hazard, as well as a source of pleasure and conviviality. Alcohol in excess can damage any part of the nervous system. The brain is vulnerable in a number of ways. Alcohol can contribute to a global dementia and can provoke, along with vitamin deficiency, the catastrophic Wernicke–Korsakoff syndrome. Alcohol can cause localized damage to the cerebellum (alcoholic cerebellar degeneration), to the pons (central pontine myelinolysis), and the corpus callosum (Marchiafava–Bignami disease). Society has used alcohol wisely from time immemorial, but newer recreational drugs have emerged, each with their own hazards. Opium, heroin, cocaine, and crack are, of course, well-known. Ecstasy is the latest to hit the headlines. 'Raves' in which ferocious hot dancing for hours without adequate fluid intake, enhanced by drugs, can rarely cause heat stroke with brain swelling and death.

All these are obvious environmental hazards, but there are other brain diseases in which the impact of the environment is less evident, but in which detective work has tracked down some clues.

PARKINSON'S DISEASE

Parkinson's disease probably affects more than 100 000 people in the United Kingdom, and costs the Government more than £1 million a year in medical and nursing care, as well as £1000 a year for each patient's medications. Overall, we may spend as much as £2 million each year for Parkinson's disease, without considering the indirect costs of lost income, missed opportunities, and family disruptions.

There is no cure for the disease, but modern drug treatment provides acceptable if temporary relief. Patients live longer, but at the price of developing more disabling and less treatable symptoms (Marsden 1994).

At the time of diagnosis, every patient probably has had Parkinson's disease for at least 4 years. It is estimated that there is at least one person undiagnosed for every patient who is recognized. The frequency of the disease increases exponentially over the age of 50 years and some 2% of the population aged 60 and over, and 4% of the population aged 70 and over, develop Parkinson's disease (Mutch *et al.* 1986). However Parkinson's disease can strike in younger individuals, and 10% of all patients are less than 40 years old at onset. The disease has even been recorded in children and adolescents.

The typical manifestations of Parkinson's disease are tremor, hence the name *paralysis agitans*, stiffness of muscles known as rigidity, slowness and hesitation of movement, and postural instability. The key pathology in the brain is degeneration of a particular nucleus in the brainstem that contains a melanin pigment, the substantia nigra, which uses dopamine as its neurotransmitter. Loss of dopamine in the basal ganglia is the cause of the motor difficulties of Parkinson's disease. Treatment is aimed at replenishing such dopamine deficiency by the administration of its precursor levodopa, or synthetic directly-acting dopamine agonists.

The cause of Parkinson's disease has been a mystery (Burton and Calne 1990). About 10–15% of patients give a family history of a similar disease, which has always raised the possibility that there may be an inherited component to the illness. However, the genetic hypothesis was discarded by many when a twin study showed similar concordance rates between monozygotic and dizygotic twins. Identical twins, one of whom had Parkinson's disease, appeared no more likely to share the illness than non-identical twins (Ward *et al.* 1983). However, subsequent reflection and re-analysis of the data has led to the conclusion that the number of twins studied was inadequate to prove or disprove the genetic hypothesis. Several large families with autosomal-dominant inherited Parkinson's disease have been described indicating that the illness could be genetically determined in some cases. Review of familial Parkinson's disease has led to the conclusion that there may well be an inherited predisposition to the illness. The similar incidence of the disease in parents and the siblings of index cases suggests that an autosomal-dominant gene (or genes) with reduced penetrance is the most likely explanation for such a pre-disposition.

At about the same time that the original twin study was published, a remarkable observation focused attention on environmental causes for Parkinson's disease. In 1982, Langston and colleagues identified a mini-epidemic of parkinsonism amongst drug addicts around San Jose in California (Langston *et al.* 1983). Inspired detective work established that all the affected young individuals had bought a new heroin substitute, made by one of the back-street chemists in operation at that time. Such 'designer drugs' could look like heroin, taste like heroin, and act like heroin, but because their chemical formulae were new, they were beyond the law (Langston and Palfreman 1995).

It turned out that another similar patient had developed young-onset parkinsonism earlier and had been admitted to the National Institutes of Health in the USA. The individual concerned was a young college student who used a chemistry set, a present from his parents, to make his own drugs. After some library research he decided to synthesize a 'designer drug' known as 1-methyl-4-phenyl-proprionoxy-piperidine, or MPPP for short. For several months he successfully made MPPP and used it himself. One day, however, he hurried a batch and then, dramatically within three days, developed severe parkinsonism. A team of scientists at NIH were not slow to see the implications of this tragedy for medical research. They reasoned that sloppy chemistry must have led to the production of some unusual substance. They managed to obtain his glassware and analysed the residue. They found that this contained not only MPPP but also 1-methyl-4-phenyl-1,2,3,6-tetrahydropyridine, or MPTP.

Sanford Markey injected MPPP and MPTP into laboratory rats, but no permanent damage ensued. The trail went cold. As it turned out, rats are not particularly susceptible to MPTP poisoning.

Langston tracked down the original paper describing this single case and contacted the NIH team. He had managed to obtain a sample of the street 'designer drug' heroin substitute that his patients in California had taken and supplied this to Ian Irwin, the Director of Stanford's Drug Assay Laboratory. Irwin was able to prove beyond doubt that the chemical present in significant amounts in the California designer heroin was in fact MPTP. Subsequently, research workers in different centres in the world established that the administration of MPTP to monkeys rapidly produced a dramatic parkinsonism more or less identical to human Parkinson's disease (Burns *et al.* 1983).

This remarkable discovery prompted a surge of scientific activity and research throughout the world, to establish how MPTP caused neurotoxicity (Langston and Irwin 1986). It was found that this agent, administered systemically, produced selective destruction of the pigmented substantia nigra in monkeys, causing neuropathological and neurochemical changes very similar to those of Parkinson's disease itself. MPTP was discovered not to be the active neurotoxin; it had to be converted via an intermediate (MPDP) by monoamine oxidase B, perhaps in glia, into 1-methyl-4- phenylpyridine (MPP^+). It was then discovered that MPP^+ is a substrate for the dopamine neuronal uptake system, which normally inactivates dopamine released into the synaptic cleft of nigrostriatal terminals. MPP^+ is thus concentrated into dopaminergic neurons, where it binds to neuromelanin – a by-product of dopamine synthesis found in nigral nerve cells, particularly in primates and in greatest concentration in man. In addition, MPP^+, once inside nigral dopaminergic neurons, is actively taken up into mitochondria.

How does MPP^+ kill dopaminergic neurons? Two complimentary mechanisms have been proposed. Most likely, it interferes with mitochondrial energy metabolism, by poisoning NADH-linked components of Complex I, leading to depletion of ATP and alterations in cellular calcium content. Alternatively, MPP^+ synthesis from MPTP may also induce the formation of free radical species imposing oxidative stress. These two mechanisms are intimately linked.

These discoveries concentrated attention on a possible environmental agent, such as MPTP, as the cause of Parkinson's disease. Indeed, a number of synthetic chemists who had worked with MPTP were discovered to have developed early-onset Parkinson's disease, but most people with Parkinson's disease had never seen or handled MPTP in its pure form. However, MPTP is a very simple molecule, a pyridine, so called because of its ring structure. Pyridines are very common in the environment; for example they are found in coal, coal tar, asphalt, some vitamins, nicotine, and many plants. Remarkably, a number of drug companies had actually looked at MPTP as a possible treatment for Parkinson's disease and depression. However, they had abandoned such work when it was discovered to be toxic to animals.

Scientists began investigating lifestyle, location, occupation, exposure to well-water, pesticides, herbicides, industrialization, and many other factors, seeking a clue to a toxic cause for Parkinson's disease in the environment (Tanner and Langston 1990; Rajput 1993). Such epidemiological research is rather like hunting for a needle in a haystack, but clues began to emerge.

Although Parkinson's disease occurs worldwide, its incidence is lower in China and Africa compared with Europe and North America. Some studies suggest that rural living, pesticide exposure, and drinking well-water might predispose to Parkinson's disease, but other studies did not. So far the evidence does not point to any single environmental cause.

The final twist in the MPTP story goes back to the unfortunate original young individuals who developed parkinsonism in California. Although they responded to conventional treatment for Parkinson's disease, they soon ran into major complications with side-effects consisting of violent abnormal movements, and a wildly fluctuating response to treatment. All attempts at drug manipulation failed to help, so eventually two of these individuals were flown to Lund, Sweden for foetal transplant surgery which has given them remarkable improvement (Widner *et al.* 1992).

THE AMYOTROPHIC LATERAL SCLEROSIS AND THE PARKINSONISM–DEMENTIA COMPLEX OF GUAM

For more than 150 years, the Chamorro people of the Mariana Islands in the Western Pacific Ocean have developed a fatal paralysis in middle and later life, termed the amyotrophic lateral sclerosis/parkinsonism dementia complex. In 1954, Drs Leonard Kurland and Donald Mulder confirmed that amyotrophic lateral sclerosis in Guam was indistinguishable from the classical form of that disease elsewhere in the rest of the world (Kurland and Mulder 1954). Seven years later Drs Asao Hirano and Kurland characterized the second neurological syndrome as parkinsonism–dementia in these people (Hirano *et al.* 1961). The two conditions, amyotrophic lateral sclerosis (otherwise known as motor neuron disease), and parkinsonism–dementia, often occurred in the same family and occasionally in the same individual.

The original observations were by navy physicians assigned to Guam, who drew the attention of the Neurological Institute of the National Institutes of Health to this epidemic. Kurland went to Guam in 1953 to investigate this remarkable phenomenon. He and his colleagues found that amyotrophic lateral sclerosis was one hundred times more common there than elsewhere in the world and that in certain villages of Southern Guam and the adjacent island of Rota, its incidence was even higher. However, it was uncommon in Saipan and did not seem to affect residents of other Mariana or Caroline Islands. More than one third of affected individuals had other family members who suffered from the same disease.

Amyotrophic lateral sclerosis patients developed muscular atrophy with weakness and fasciculation, associated with pyramidal signs, affecting bulbar and limb muscles. Patients generally presented in their 40s or 50s. The disease was relentlessly progressive with death occurring within 5 to 10 years. The parkinsonism–dementia syndrome was characterized by severe akinesia with rigidity and a progressive loss of higher intellectual function. The disorder usually began later, in the sixth or seventh decade, but also was progressive with death within 4–6 years. The pathology of the condition consists of severe neuronal loss in the cerebral cortex and brainstem pigmented nuclei, associated with neurofibrillary tangles in remaining nerve cells (Rodgers-Johnson *et al.* 1986).

The high incidence of the illness amongst the Chamorro people in Guam and Rota, compared to the very low incidence in the Chamorro of the same genotypic origins in Saipan and other Northern islands, strongly suggested that this familial disease was due to environmental factors. Other observations pointed to the same conclusion. Filipino migrants were found to develop the disease after long-term residence, usually more than 13 years, on Guam. Chamorro migrants to the United States mainland retained their increased risk of developing the disease as long as three decades after migration, if they had had at least 18 years of exposure to the Guamanian environment (Steele and Guzman 1987).

In a remarkable piece of detective work, the likely environmental toxin responsible for this illness has been identified as cycad seeds (Kurland 1972). The illness occurs only in those islands where cycad trees grow. Guam and Rota contain dense cycad forests. In the early nineteenth century, poverty was widespread and food resources declined so that the Chamorro were often forced to scavenge in the forests for food. It was about this time that the native custom of preparing flour from cycad seeds developed. The indigenous cycad, *Cycas circinalis*, survived typhoons when all other food sources were destroyed. The seeds were known to be poisonous, but Spanish speaking immigrants taught the natives to soak them first to remove the poison. They were then sun-dried and could be stored for many months before being ground into a fine flour which was used in the preparation of tortillas. Though the use of cycad as a food source began as a necessity during a time of privation, it subsequently became a dietary staple for many families in Guam. Spencer and his associates identified a toxin of the cycad seed, β-*N*-methylamino-L-alanine (BMAA), which they showed could cause damage to the motor system of monkeys (Spencer *et al.* 1986).

After World War II, considerable quantities of food were brought to the islands, including wheat flour which replaced the use of cycad flour. This change in food habit occurred 40 years ago. As a result, the incidence of amyotrophic lateral sclerosis and parkinsonism–dementia began to decline in the 1950s and 60s, such that they have now virtually disappeared.

PRION DISEASES

The next example is the remarkable story of an infectious agent in the environment which causes devastating destruction of the brain, and which has generated great recent interest in the form of 'mad cow disease'.

The story began with sheep. As early as 1738, the disease 'scrapie' was recognized amongst sheep and goats. In 1938, it was demonstrated that scrapie could be transferred experimentally from one sheep to another, arguing for an infectious aetiology. Meanwhile, it was also realized that the genetic background of flocks profoundly influenced their susceptibility to scrapie, raising the possibility that it might be an inherited disorder. These two opposing views sparked many controversial encounters and set the scene for what subsequently developed (Parry 1962; Dickinson *et al.*, 1965).

For many decades, a disease known as Kuru devastated the lives of the Fore Highlanders of Papua New Guinea. The illness began with difficulty in walking which became progressively more severe with the emergence of a profound cerebellar ataxia. Most patients died within a year of onset, and all were dead within two years. Curiously, the illness occurred mainly among women, and at the height of the epidemic it was unusual in the Fore region to see an old woman. Remarkable detective work by Carlton Gajdusek established that the disease was caused by ritualistic cannibalism (Gajdusek 1977). Women and children were given the offal, including the brain, while men ate meat.

In 1966, Gajdusek and colleagues transmitted Kuru to chimpanzees experimentally using brain inoculation, establishing it as an infectious disease (Gajdusek *et al.* 1966). The incubation period was very long, so the idea of a slow virus infection developed. Cannibalism ceased by 1960 in the Fore region, and no individual born in that area after 1959 has developed Kuru. The incubation period may be as long as three decades, but Kuru has progressively disappeared, first among children and thereafter among adolescents. The number of deaths in adult women has decreased steadily and each year the youngest new patients are older than those in the previous year.

The neuropathological changes in Kuru are striking. The brain becomes spongy, there is marked astrogliosis and amyloid or Kuru plaques are distinctive. This spongiform degeneration was recognized to be similar to that of scrapie, and to another human brain disease, namely Creutzfeldt–Jakob disease. This observation led Gajdusek to attempt transmission of Creutzfeldt–Jakob disease to apes which was successfully achieved in 1968 (Gibbs *et al.* 1968).

Creutzfeldt–Jakob disease is found throughout the world, but is extremely rare with approximately one case per million of the population. Most cases occur sporadically, but 10–15% of cases are familial when it is inherited as an autosomal-dominant trait with variable penetrance. The illness causes a rapid and devastating dementia leading to death within months in the majority (Masters and Richardson 1978; Brown *et al.* 1986).

Over 250 cases of Creutzfeldt–Jakob disease have now been transmitted to apes and monkeys, clearly indicating that it is due to an infective agent. The incubation time ranges from 4 months to more than 4 years. There also have been a number of examples of iatrogenic transmission of Creutzfeldt–Jakob disease as a result of corneal transplantation, contaminated EEG electrode implantation, and surgical operations using contaminated instruments or apparatus. Creutzfeldt–Jakob disease has also been produced by implantation of dura mater grafts from infected hosts (Davanipour *et al.* 1984).

The largest outbreak of iatrogenic Creutzfeldt–Jakob disease has occurred as a result of transmission from contaminated growth hormone preparations derived from human pituitaries, one or more of which must have come from an individual harbouring the illness (Brown *et al.* 1992). Such human growth hormone was used to treat small stature in children, and more than 40 patients ranging in age from 10 to 41 so treated have developed Creutzfeldt–Jakob disease. The incubation period has been estimated to be between 4 and 30 years. How many lots of human growth hormone might have been contaminated with the infectious agent is not known.

All this evidence strongly points to the conclusion that Creutzfeldt–Jakob disease is due to an infectious agent, albeit one with a very long incubation period. There remains the problem of explaining the 10–15% of cases which are familial and inherited. In the context, another disease emerged of importance. Gerstmann-Sträussler Schenker disease was known to be a familial condition inherited as an autosomal-dominant trait (Kuzuhara *et al.* 1983). It is rare, comprising less than 2% of all Creutzfeldt–Jakob or Creutzfeldt–Jakob related diseases. Patients develop a variety of symptoms, but typically there is a progressive cerebellar ataxia with onset in adult life, leading to death within about 5 years. Pathologically, the brain shows spongiform degeneration with amyloid plaques similar to that seen in Creutzfeldt–Jakob disease and Kuru. In 1978 Tateishi and co-workers transmitted Gerstmann–Sträussler–Schenker disease to animals (Tateishi *et al.* 1979). So here was an apparent hereditary disease which also was infectious.

The nature of the infectious agent causing these various spongiform degenerations of the brain has always been controversial. Prusiner has established that it is a novel infectious agent, different from classical viruses or viroids (Prusiner 1994). In 1982 Prusiner introduced the term '*prion*' to describe a small proteinaceous infectious particle which resists inactivation by procedures that modify nuclear acids (Prusiner 1982). The brain normally contains prion proteins (PrP^{c}), whose function is unknown, but which are not infectious. However, infectious and transmissible prions consist of an abnormal isoform of prion protein designated PrP^{sc}. The normal PrP is transformed by a post-translational process into the infective PrP^{sc} or scrapie agent by an unknown mechanism. Accumulation of PrP^{sc} causes the brain disease, but how it replicates is unknown (Prusiner 1994).

Thus, the picture emerges of a normal prion protein coded for by the prion protein gene on chromosome 20, the function of which is unknown. Deletion of the prion protein gene in young mice seems to have no deleterious effect. Mutations of the prion protein gene are responsible for familial Creutzfeldt–Jakob disease, Gerstmann–Sträussler–Schenker disease, and Fatal Familial Insomnia. Eighteen different mutations in the human PrP gene have been identified so far, all linked genetically or segregating with these inherited human prion diseases. Such mutations lead to accumulation of the abnormal form of the prion protein, the scrapie isoform, in the brain. All these familial conditions are transmissible to other animals, particularly primates, by brain inoculation. Thus these remarkable illnesses are both genetic and infectious.

The abnormal scrapie isoform of the prion protein also appears to be the infectious agent responsible for Kuru and most cases of sporadic Creutzfeldt–Jakob disease. Perhaps the scrapie isoform represents a somatic mutation of the prion protein gene. The infectious and transmissible abnormal prion is quite different from any known virus or viroid; infections have a long incubation period lasting years to decades, but when these illnesses strike they prove rapidly fatal.

All this is relevant to 'mad cow disease' (Wilesmith and Wells 1991). Since 1986 more than 100 000 cattle have died of 'mad cow disease' (bovine spongiform encephalopathy) in Great Britain. Many investigators contend that bovine spongiform encephalopathy was the result of feeding cattle on meat and bone meal prepared from rendered sheep offal, assumed to be harbouring scrapie. A change in the method of extracting sheep offal prior to the appearance of bovine spongiform encephalopathy has been suggested as the reason why scrapie prions survive the rendering process (Spongiform Encephalopathy Advisory Committee 1994). Since 1988, feeding rendered meat and bone meal has been banned, and the number of cases of bovine spongiform encephalopathy has begun to decrease (see Roberts 1995).

The fear is that the prion agent responsible for bovine spongiform encephalopathy will cross the species barrier into man. An outbreak of transmissible mink encephalopathy in 1985 has been said to be the consequence of feeding mink on meat and bone meal from a cow with a sporadic case of bovine spongiform encephalopathy. Likewise, there have been a number of cases of feline spongiform encephalopathy amongst cats (also affecting a puma and a cheetah) and exotic ungulate encephalopathy in zoos (affecting a nyala and a greater kudu), attributed to feeding infected beef offal.

Whether bovine spongiform encephalopathy poses any risk to humans is at present unknown. Careful surveillance on the incidence of Creutzfeldt–Jakob disease in the United Kingdom has suggested that the number of cases, some 55 per year, may have increased recently (see Hofman and Wientjens 1995). However, a similar increase has been evident in countries in which bovine spongiform encephalopathy does not occur, so may be attributable to a greater awareness and ascertainment of the disease. Nevertheless, there are obvious concerns that the outbreak of bovine spongiform encephalopathy in cattle, entering the human food chain in the form of infected meat and other products, might after a long incubation period lead to an outbreak of Creutzfeldt–Jakob or similar diseases in humans in the future. Such fears have promoted much speculation in the media, and political stances over the safety or otherwise of British beef. Some have changed their eating habits, while others vigorously defend the wholesomeness of burgers and steaks. A nationwide system for the surveillance of the incidence of Creutzfeldt–Jacob disease has been put in place in the United Kingdom, and Europe. No definite trend has emerged as yet, but the jury is still out (see *British Medical Journal*, November 1995, **311**, 1415–21 for a contemporary survey of differing views).

CONCLUSIONS

The snapshots presented here illustrate the range of environmental risks to the human brain. Trauma, toxins, infections, and genetic factors pose threats to the organ that is the repository of intelligence, thought, communication, and wisdom. Perhaps the greatest challenge is to understand the origins of those degenerative diseases that affect the aging human brain. If we are to live longer, we must hope that we can do so as intelligent humans who can continue to contribute to society. But the longer we live, the greater the risks of disintegration of brain function. The discovery of the interaction of environmental and genetic hazards to brain function is a formidable, but potentially rewarding goal for neuroscientific research.

REFERENCES

Brown, P., Cathala, F., Castaigne, P., and Gajdusek, D.C. (1986). Creutzfeldt–Jacob disease: clinical analysis of a consecutive series of 230 neuropathologically verified cases. *Annals of Neurology*, **20**, 597–602.

Brown, P., Preece, M.A., and Will, R.G. (1992). 'Friendly fire' in medicine: hormones, homografts, and Creutzfeldt–Jacob disease. *Lancet*, **340**, 24–7.

Burns, R.S., Chiueh, C.C., Markey, S.P., Ebert, M.H., Jacobowitz, D.M., Kopin, D.M. (1983). A primate model of parkinsonism: selective destruction of dopaminergic neurons in the pars compacta of the substantia nigra by N-methyl-4-phenyl-1,2,3,6-tetrahydropyridine. *Proceedings of the American Academy of Science*, **80**, 4546–50.

Burton, K. and Calne, D.B. (1990). Aetiology in Parkinson's disease. In *Parkinson's disease* (ed. G. Stern), pp. 269–93. Chapman and Hall Medical, London.

Davanipour, Z., Goodman, L., Alter, M., Sobel, E., Asher, D., Gajdusek, D. C. (1984). Possible modes of transmission of Creutzfeldt–Jacob disease. *New England Journal of Medicine*, **311**, 1582–3.

Dickinson, A.G., Young, G.B., Stamp, J.T., and Renwick, C.C. (1965). An analysis of natural scrapie in Suffolk sheep. *Heredity*, **20**, 485–503.

Gajdusek, D.C. (1977). Unconventional viruses and the origin and disappearance of kuru. *Science*, **197**, 943–60.

Gajdusek, D.C., Gibbs, C.J.Jr., and Alpers, M. (1966). Experimental transmission of a kuru-like syndrome to chimpanzees. *Nature*, **209**, 794–6.

Gibbs, C.J.Jr., Gajdusek, D.C., Asher, D.M., Alpers, M.P., Beck, E., Daniel, P.M., *et al.* (1968). Creutzfeldt–Jacob disease (spongiform encephalopathy): transmission to the chimpanzee. *Science*, **161**, 388–9.

Grey, A. and Fenn, P. (1993). Alzheimer's disease: the burden of the illness in England. *Health Trends*, **25**, 31–7.

Hirano, A., Kurland, L.T., Krooth, R.S., and Lessel, S. (1961) Parkinsonism-dementia complex, an endemic disease on the island of Guam. *Brain*, **84**, 642–79.

Hofman, A. and Wientjens, D.P.W.M. (1995). Creutzfeldt–Jacob disease and bovine spongiform encephalopathy: any connection? Epidemiological evidence concerning a possible causal link. *British Medical Journal*, **311**, 1418–19.

Jennett, W.B. and MacMillan, R. (1981). Epidemiology of head injury. *British Medical Journal*, **282**, 101–7.

Jorm, A.F. (1990). *The epidemiology of Alzheimer's disease and related disorders*. Chapman and Hall, London.

Khachaturian, Z.S. (1985). Diagnosis of Alzheimer's disease. *Archives of Neurology*, **42**, 1097–105.

Kurland, L.T. (1972). An appraisal of the neurotoxicity of cycad and the etiology of amyotrophic lateral sclerosis of Guam. *Federation Proceedings*, **31** (5), 1540–2.

Kurland, L.T. and Mulder, D.W. (1954). Epidemiological investigations of amyotrophic lateral sclerosis: 1. Preliminary report on geographic distribution, with special reference to the Mariana Islands, including clinical and pathological observations. *Neurology*, **4**, 355–78; **5**, 438–48.

Kuzuhara, S., Kanazawa, I., Sasaki, H., Nakanishi, T., and Shimamura, K. (1983). Gerstmann–Sträussler–Scheinker's disease. *Annals of Neurology*, **14**, 216–25.

Langston, J.W. and Palfreman, J. (1995). *The case of the frozen addict*. Pantheon Books, New York.

Langston, J.W., Ballard, P., Tetrud, J.W., and Irwin, I. (1983). Chronic parkinsonism in humans due to a product of meperidine analog synthesis. *Science*, **219**, 979–80.

Langston J.W. and Irwin I. (1986). MPTP: current concepts and controversies. *Clinical Neuropharmacology*, **2** (6), 485–507.

Marsden, C.D. (1994). Parkinson's disease. *Journal of Neurology, Neurosurgery and Psychiatry*, **57**, 672–6.

Masters, C.L. and Richardson, E.P.Jr. (1978). Subacute spongiform encephalopathy Creutzfeldt–Jacob disease—the nature and progression of spongiform change. *Brain*, **101**, 333–44.

Miller, J.D. and Jones, P.A. (1985). The work of a regional head injury service. *Lancet*, **1**, 1141–4.

Mutch, W.J., Dingwall-Fordyce, J., Downie, A.W., Paterson, J. G., and Roy, S. K. (1986). Parkinson's disease in a Scottish city. *British Medical Journal*, **292**, 534–6.

Office of Technology Assessment, US Congress. (1987). Losing a million minds: confronting the tragedy of Alzheimer's disease and other dementias. Washington DC: US Government Printing Office, Publ No. OTA-BA-323.

Parry, H.B. (1962). Scrapie: a transmissible and hereditary disease of sheep. *Heredity*, **17**, 75–105.

Prusiner, S.B. (1982). Novel proteinaceous infectious particles cause scrapie. *Science,* **216**, 136–44.

Prusiner, S.B. (1994). Prion diseases of humans and animals. *Journal of the Royal College of Physicians of London*, **28**, Supplement No. 2, 1–30.

Rajput, A.H. (1993). Environmental causation of Parkinson's disease. *Archives of Neurology*, **50**, 651–2.

Roberts, G.W. (1995). Creutzfeldt–Jacob disease and bovine spongiform

encephalopathy: any connection? Furrowed brow over mad cow. *British Medical Journal*, **311**, 1419–20.
Rodgers-Johnson, P., Garruto, R.M., Yanagihara, R., Chen, K.-M., Gajdusek, D. C., and Gibbs, C.J. Jr. (1986). Amyotrophic lateral sclerosis and parkinsonism–dementia on Guam: a 30-year evaluation of clinical and neuropathologic trends. *Neurology*, **36**, 7–13.
Rossor, M.N. (1995). Dementia. In *Management of neurological disorders* (ed. C.M. Wiles), pp. 204–9. BMJ Publishing Group, London.
Skoog, I., Nilsson, L., Palmertz, B.O., Andreasson, L-A., and Svanborg, A. (1993). A population-based study of dementia in 85-year-olds. *New England Journal of Medicine*, **328**, 153–8.
Spencer, P.S., Nunn, P.B, Hugon, J., Ludolph, A., and Dwijendra, N. R. (1986). Motor neuron disease of Guam: possible role of a food neurotoxin. *Lancet*, **1**, 965.
Spongiform Encephalopathy Advisory Committee. (1994). *Transmissible spongiform encephalopathies: a summary of present knowledge and research*. HMSO.
Steele, J.C. and Guzman, T. (1987). Observations about amyotrophic lateral sclerosis and the parkinsonism-dementia complex of Guam with regard to epidemiology and etiology. *Canadian Journal of Neurological Sciences*, **14**, 358–62.
Strang, I., MacMillan R., and Jennett, B. (1987). Head injuries in accident and emergency departments at Scottish Hospitals. *Injury*, **10**, 154–9.
Tanner, C.M. and Langston, J.W. (1990). Do environmental toxins cause Parkinson's disease: a critical view. *Neurology*, **40**, (suppl 3), 17–30.
Tateishi, J., Ohta, M., Koga, M., Sato, Y., and Kuroiwa, Y. (1979). Transmission of chronic spongiform encephalopathy with kuru plaques from humans to small rodents. *Annals of Neurology*, **5**, 581–4.
Teasdale, G.M. (1995). Head injury. In *Management of neurological disorders* (ed. C.M. Wiles), pp. 88–125. BMJ Publishing Group, London.
Ward, C.D., Duvoisin, R.C., Ince, S.E., Nutt, J.D., Eldridge R., and Calne, D.B. (1983). Parkinson's disease in 65 pairs of twins and in a set of quadruplets. *Neurology*, **33**, 815–24.
Widner, H., Tetrud, J., Rehncrona, S., Snow, B., Brundin, P., Gustavi, B. *et al.* (1992). Bilateral fetal mesencephalic grafting in two patients with parkinsonism induced by MPTP. *New England Journal of Medicine*, **327**, 1556–63.
Whitehouse, P.J. (1993). *Dementia*. F.A. Davies Co, Philadelphia.
Wilesmith, J. and Wells, G.A.H. (1991). Bovine spongiform encephalopathy. *Current Topics in Microbiology and Immunology*, **172**, 21–38.

6

Mental health and the urban environment

Hugh Freeman

Professor Hugh Freeman, DM, FRCPsych, has been Honorary Consultant Psychiatrist to the Salford Health Authority and to the Medical School of Manchester University since 1988. He took his first degree, in medicine, at the University of Oxford and was subsequently commissioned in the Royal Army Medical Corps. He then held appointments as Registrar at the Bethlem Royal and Maudsley Hospitals in London and at Littlemore Hospital in Oxford before taking up the post of Consultant Psychiatrist at Salford Royal Hospital. This began a long association with the Departments of Health and Social Services in Salford which continued until 1988. Professor Freeman has chaired or served on numerous public committees, panels, and working parties in the fields of mental health and psychiatry and is the author of a wide range of publications in these fields. He was Editor of the British Journal of Psychiatry *and of* Current Opinion in Psychiatry*; he edited and contributed to* Mental Health and the Environment, *which was published in 1985.*

The relationship between mental health and the environment is a surprisingly neglected subject in spite of the greatly increased importance of the environment as a political and scientific issue in recent years. As a result, it is a scientific area that still lacks both adequate data and rigorously defined concepts. The problem was well expressed by the great environmentalist, René Dubos (1972): 'The study of man as an integrated unit and of the ecosystems in which he functions is grossly neglected because it is not in the tradition which has dominated science since the seventeenth century'. As a result, there is now an urgent need to examine those aspects of people's physical and social surroundings which are likely to influence mental health.

This chapter focuses on the urban macro-environment rather than micro-environments, such as the family and internal environments of institutions such as hospitals. For the first time in history, half the world's population now live in cities. In the forty years between 1950 and 1990, the population of the world's cities increased tenfold. The urban population of the world is increasing at a rate of a quarter of a million a day. In Bombay, five million people live in shanties or on the streets – equal to the entire population of inner London.

CAUSES OF PSYCHIATRIC DISORDERS

Psychiatry generally operates on the basis of a multi-factorial model of causation, in which environmental factors are assumed to interact with biological and other influences. Within this model, whatever contribution is not genetic, such as exposure to a certain type of housing, must be environmental. But to take such an extreme reductionist view would not be scientifically useful, since there are a number of different ways in which genetic – environmental interactions can occur, and some of the component causes will always be unknown. For several reasons, direct cause-and-effect relationships between specific features of the environment and abnormal mental states will not be easy to find.

Within any one person, the intervening processes between an environmental stimulus and an eventual health outcome must be enormously complex. These processes are located primarily in the central nervous system and the endocrine system. Depending on the particular circumstances, the outcome might be a psychiatric disorder, a physical disorder, or no abnormality at all, as when moderate stress improves performance. Secondly, any one feature of the environment can have very varying effects on different people who are exposed to it, because of their different genetic endowments and life experiences. Thirdly, 'mental health' should not be treated as a single variable, when it is actually a global term that is often used in a very vague way. If any environmental factor does have a pathogenic effect, this may be only in relation to one major category of psychiatric disorder, such as schizophrenia. It would mostly be unsafe to generalize from one category of illness to another. Finally, the urban environment needs to be deconstructed into factors which can be studied separately, if any meaningful effects are to be found. In fact, health – whether physical or mental – is only one of the factors used by any society to determine whether an environment is considered 'good' or socially acceptable. Where a choice is available between one environment and another, economic, cultural, or even aesthetic factors may be at least as important as health.

Both in Europe and North America, the rapid growth of cities in the nineteenth century caused tremendous overcrowding, associated with extreme poverty and the ravages of infectious diseases such as cholera. Working-class housing generally did not provide enough space for a reasonable quality of life, particularly for the large families that were then common. As a result, planners and architects became preoccupied with the reduction of overall urban density as a primary goal, but in this search, they overlooked the important difference between density and household crowding (see pp. 136–7).

In this discussion, the relationship between the physical environment and human behaviour is a crucial issue. Sociological work has shown that when aspects of the physical environment are significant in people's lives, they become part of these people's social norms – of the way they use their environment in their daily lives (Gans 1967). These social norms in turn can have either positive or negative effects on people's well-being. But before there can be an effect on individuals, environmental influences are usually mediated through both social and cultural factors, which include mental states. For instance, a depressed person will probably see the same environment in a much more negative way than a non-depressed person. As a result, the meaning that is given to any environmental factor depends on the characteristics of the people concerned. This 'response bias' is very likely to affect efforts to establish causality in relation to environmental factors: 'Where studies have not included objective independent measures of the environmental variables, the causal directions of the associations are difficult to establish' (Halpern 1995).

INVESTIGATION OF RELATIONSHIPS

Two conceptual tools have been used in investigating the relationship between environmental variables on the one hand and the characteristics of individuals or communities on the other.

The first of these is social support, which is related to the theory of social networks, but there are considerable methodological problems in using this conceptual tool. For instance, does poor social support cause psychiatric disorder, or is it the other way round? What is the direction of causality? There is reasonably firm evidence that social support is an important factor in depression (Brown and Harris 1978), in many neurotic disorders, and in buffering the effects of stress. Sources of support include the human interactions that take place in cafés or pubs, or in the outdoor *paseo* of Mediterranean countries. Other environmental features, such as a street fountain, may become a focus of social interaction, but this will only be possible if there is no traffic surrounding it.

In more dispersed, motorized societies, informal, unplanned social support of this kind becomes less and less possible. The endless spread of housing around large cities requires more and more car travel for any activity. Out-of-town shopping centres are so large and draw on such a huge catchment area that the likelihood of seeing anyone that one knows is considerably less than in a local high street. Yet the possibility of chance meetings and face-to-face encounters is a fundamental part of human life. A vibrant, diverse, and densely populated city is the environment that fosters these interactions in the best way (Sherlock 1991). The Inns of Court in London represent an outstanding example of this, which have adjusted to present-day technological communications.

The effects of commuting have been little studied, but there are strong prima facie grounds for proposing that this is often very harmful – to the commuter, the family, and society in general. The residential areas become dormitories which are largely devoid of activity during weekdays, while the business districts are deserted in the evenings and at weekends; in both cases, social support is absent for much of the time. The intermediate areas also suffer because of the volumes of traffic which pass through them (Wedmore and Freeman 1984). Negative environmental effects can be found in the perceived danger which causes people to avoid certain places. Their avoidance then increases the danger, because fewer people are about and isolated individuals are more vulnerable. In this situation, chronic anxiety disorders are likely to develop, particularly among older people.

For example, in the redeveloped central area of the northern (UK) city of Salford, built in the early 1970s, the planners provided a large open space, which probably looked good in the drawings, but this space has no function (Fig. 6.1). People will cross it as quickly as they can, particularly in the wet weather which is very frequent, and they may well see it as dangerous at certain times. In these ways, structural environments can either increase or decrease social support, which in turn is likely to have some effects on mental health.

The second conceptual model used in analysing these relationships is that of stress. Psychiatry has a long tradition of regarding external stress as a causal factor in mental disorder. In early mental hospitals, such as The Retreat at York (Digby 1985), a withdrawal from social responsibilities, a benevolent but firm regime, and an undemanding occupation were believed to promote a healing process. In today's terms, it would be seen as protection from environmental stress.

There are both physiological and psychological aspects to stress; each may be associated both with direct harmful effects on the individual and with the negative consequences of coping. But hardly any direct relationships have been found between particular forms of environmental stress and specific psychiatric disorders. An important exception is the post-traumatic stress disorder which follows environmental disasters – whether natural, such as Australian bush fires or man-made, such as Chernobyl or Bhopal in India. The Athens earthquake of 1981 was in fact found to have been associated with

Fig. 6.1. Central redeveloped area of Salford.

a measurable epidemic of heart attacks, a few days later, providing a remarkable natural experiment of the effects of stress (Trichopoulos *et al.* 1983).

'Stress' is a poorly defined concept, and its explanatory power is limited, but so far, there seems to be no alternative to using it. In relation to the environment, there are several different forms which need to be considered.

'Chronic stress' refers to adverse factors which endure over time and from which there is little possibility of escape. Those chronic stresses which come from environmental conditions may be particularly difficult to change, and there is some evidence that these may be involved in the causes of psychiatric disorders (Brown 1991).

One of the commonest examples is noise, which has become very much more prevalent with the growth of motor and air traffic and the various forms of amplified sound. 'Noise' is a psychologically loaded term for a particular form or level of sound. There are marked cultural differences in what is seen as an acceptable, non-stressful level of noise input. For instance, northern Europeans are generally less tolerant of noise than Mediterranean people. One thing, though, is clear from the published research – that noise does not increase the risk of severe psychiatric disorder, such as schizophrenia. So far as less severe morbidity is concerned, noise has been shown to have very complex relationships with three independent phenomena. These are: sensitivity, annoyance, and neuroticism (or vulnerable personality). People in poorer mental health are more sensitive to noise, as well as to other stresses. A small minority of the population do not show normal habituation to noise, that is they do not get used to it with repetition so that they suffer more from annoyance. If they also have neurotic traits, the result may be an increase in less severe psychiatric disorders (Tarnopolsky and Clark 1984). Figure 6.2 models the relationship between these phenomena.

The effects of traffic noise have hardly been studied scientifically; the research carried out so far has found that it has less effect than might be expected (Halpern 1995).

Another concept often used in environmental studies is 'social stress', which proposes that adverse social factors have a role in causing psychiatric disorder. Such adverse factors include: poverty; stigmatization, (for example handicap, some forms of minority status, or living in a 'bad' area); and role disturbance, such as one-parent families or prolonged unemployment.

From the environmental point of view, an important aspect is that these adverse factors show spatial variations. They are more common in certain places, such as inner cities and peripheral public housing estates in industrialized countries and tend to cluster in either a positive or a negative direction. Where the clustering is negative, there is likely to be an association both with higher levels of psychiatric illness and of social pathology, such as drug-taking, child abuse, delinquency, and attempted suicide. There is also likely to be more pollution in such areas.

Environments of this kind, where many adverse social factors are present at a high level, have been described as showing 'socio-cultural disintegration' (Leighton 1959). Social disintegration theory assumes that the level of social structure in any population plays an important part in determining whether or not an individual in that population becomes mentally ill. This is closely allied to the concept of 'anomie', introduced into sociology by Emil Durkheim, at the end of the nineteenth century. A similar concept is 'alienation', though this has been much misused by Marxists.

Anomie is a concept which embraces the loss of both social support and of an overall moral order in society. Where the forces that bind members of a society together are weakened, people may be adversely affected, for instance in their mental health. Rapid social and cultural change will tend to cause such a weakening of bonds. These changes may include large movements of population, both through comprehensive redevelopment, and through rapid changes in the socio-economic or ethnic composition of a neighbourhood.

Durkheim claimed (1897) that the suicide rate in any country could be related to the strength of anomie there. National suicide rates do

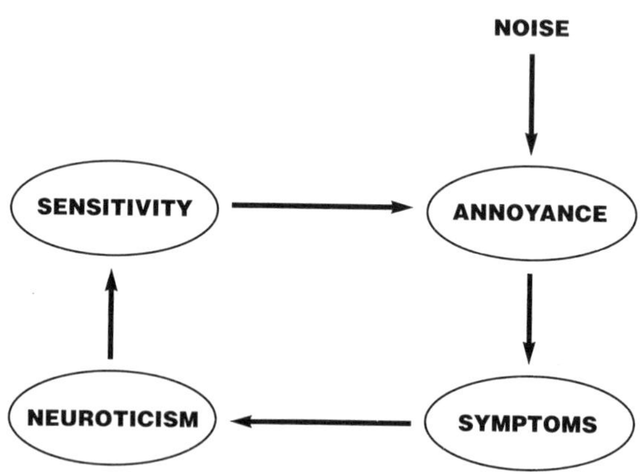

Fig. 6.2 Relationship of noise to psychiatric disorder. (Modified from Halpern 1995).

vary considerably, even when allowance is made for the often poor reliability of the statistics. They are particularly high in Northern and in Central Europe – though these two regions are very different from each other. But in itself, anomie is much too simplistic an explanation for these variations, and suicide rates also vary within countries.

Sainsbury (1955) showed that in London, suicide was most common in areas which had the highest measures of social isolation. These high-rate areas were mainly central, but not all of them showed socio-cultural disintegration, and not all were poor – for instance, parts of north-west London with many subdivided houses and blocks of flats. However, deliberate self-harm, or attempted suicide, shows even stronger spatial variations than completed suicide. Unlike suicide, it also clusters with other forms of social pathology, such as delinquency in areas of social deprivation (Buglass and Duffy 1978).

So far as housing conditions are concerned, there have been surprisingly few scientifically acceptable studies. A study of working-class women in London by Birtchnell *et al.* (1988) found that those living in poorer environments had an excess of depression. This confirmed the common-sense view that poor housing is bad for one, but the direction of causality was uncertain. Those women who were previously vulnerable to poorer mental health might have obtained poorer housing as a result of that vulnerability. This study could not resolve that question because it had not examined the women before they moved into the accommodation.

Brown's (1991) influential three-factor theory of the causes of depression includes the category, Major Difficulties. One of the most common of these difficulties is unsatisfactory housing.

Another chronic stress is air pollution. The extent of the dirty atmosphere in industrial areas up to the 1960s is almost unbelievable today, yet millions of people in the UK have lived most of their lives in a poisonous environment. However, any effects on mental health remained unknown, since they were never investigated. Although smoke from coal burning has been largely eliminated, atmospheric pollution is now almost as bad again in Britain, mainly through power station emissions and exhaust fumes from vehicles.

A Californian study (Cohen *et al.* 1986) showed that air pollution had no direct effect on mental health, but that it did act as a vulnerability factor – the effects of pollution were only evident in the simultaneous presence of a stressful life event. Thus the cost to the individual of coping with such an undesirable, polluted environment is a reduced ability to deal with new stresses. However, the methodological issues in these studies are extremely complex, and hardly any of them have been able to examine more than one environmental factor at the same time.

Another important contribution to the subject can be made from an ecological perspective. The paradigm comes from biology, and is the study of relationships between a habitat and the species which that habitat contains. In human social terms, it concerns the relationships between a local territory and the people occupying it. These relationships are found to be influenced by the characteristics of that particular territory, for instance, whether it is urban or rural.

The Chicago school of sociology, which began in the 1920s, proposed that cities can be divided into areas, or zones, which show distinctive socio-economic characteristics. These characteristics can then be related to the health and behaviour of the people living in each zone (Park and Burgess 1925). Faris and Dunham (1939) plotted the home addresses of people admitted to mental hospitals with schizophrenia from the city of Chicago. They found that the highest rate was in the most central zone of the city, and that it became steadily less towards the outer suburbs. There was no such pattern for depressive disorder or for dementia; these cases were distributed randomly throughout the city. In the central areas, there was a declining population, environmental deterioration, much casual employment, and a high rate of people living alone. Their conclusion was that these environmental conditions led to social isolation, which in turn promoted the development of schizophrenia. This was described as the 'breeder hypothesis'.

However, when the study was repeated by Levy and Rowitz (1974), and population movements were examined, an entirely different picture emerged. The overall high-rate areas, combining old and new cases, were still central, and the low-rate areas peripheral. But new cases of schizophrenia were randomly distributed throughout the city, in terms of their family homes. The central excess came from the migration of individuals who were becoming ill or had been ill, and who had deteriorated in social performance. They had become unable to continue living in better residential districts (Fig. 6.3).

This migration, which also applied to other socially marginal groups, such as Skid Row alcoholics, was described as 'social drift' (Hare 1956). In schizophrenics, it also involves a general drift downwards in the social class scale. In this way, the process tends to concentrate people who are less able and more impaired in less desirable areas. At the same time, an opposite migration was identified

– described as 'social selection' – of healthier people moving outwards to suburban areas. These are both forms of social segregation. The combined migrations produce what has been described as a 'residual' population, with high levels of morbidity, both mental and physical. Thus, being resident in a poor environmental area was not an aetiological factor in itself, but rather one associated with the adverse effects of developing schizophrenia.

Although prevalence rates of schizophrenia are generally higher in cities, the same excess can be found in some areas which have the opposite environmental characteristics. One notable example is the rural West of Ireland. There, it has been shown that the very high prevalence rate of total cases is not due to a higher-than-average incidence rate of new cases (Youssef *et al.* 1991). This excess prevalence rate in Ireland seems to be related to a number of social processes, of which the general decline in population, associated with migration, is probably the most important.

The same effect was found in epidemiological studies of Salford, where the rate was at the top of the range reported in international

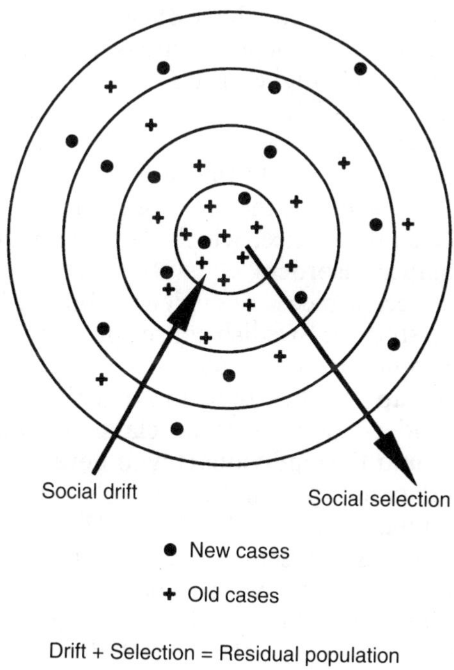

Fig. 6.3. Social segregation and schizophrenia.

studies (Freeman and Alpert 1986). After ten years, in 1984, the prevalence rate showed a further slight increase (Bamrah *et al.* 1991). The background to this finding is that in the fifty years from 1928, the inner city population had fallen to 50% of the starting-point. In the West of Ireland, the fall was of the same extent, but it began much earlier there, in the late nineteenth century.

In this connection, what is most relevant about the decline in the population of Salford and similar industrial areas is that it took a differential form. Healthier, economically active people, often in family units, were more likely to move out to better surroundings. Those left behind were more likely to have morbidity of all kinds, to be unemployed, and not to be part of a complete family. They would also tend to lose social support as the population fell – as would those remaining in the rural West of Ireland. This is the 'residual' population, mentioned above.

On the other hand, suburban areas and new towns tend to have relatively low rates of severe and chronic psychiatric disorders, reflecting the fact that people coming to settle in them are mainly younger adults with children, who are economically active and physically healthy (Taylor and Chave 1964). The policy implications of this work are that socially deprived areas, particularly of the inner cities, will contain much higher rates than average of psychiatric and social abnormalities. This means that the appropriate medical and social services need to be provided at a higher level than in more favoured areas – precisely the opposite of that mostly found – services are usually worse in the areas of high demand. This applies particularly to general practice, which deals with the great bulk of less severe psychiatric disorders.

There are three recent pieces of evidence about schizophrenia and city life. In a large study of Swedish conscripts (Lewis *et al.* 1993), the incidence of schizophrenia was 1.65 times higher among those who had been brought up in cities than those with a rural upbringing. This excess was independent of social class. In Camberwell, South London, it was found that individuals who developed schizophrenia were more likely than non-psychiatric controls to have been born into socially-deprived households (Castle *et al.* 1993). In Nottingham, there was a trend for new cases of schizophrenia to have been living in the most deprived areas of the city five years before being identified, so that they had not drifted there through illness (Dauncey *et al.* 1993).

This represents a return to the 'Breeder' hypothesis – perhaps through maternal viral infections and other hazards to the fetal nervous system in deprived social conditions. The Swedish excess, though, was not related to social deprivation, so that the overall picture is not clear. Therefore, it is not yet certain that 'urbanicity' – whatever that may be – is an independent causative factor in schizophrenia (Freeman 1994).

Is living in cities bad for mental health?

For centuries, there has been a belief that living in cities is unhealthy, because of a presumed greater social stress in urban life, as well as the very real problems of hygiene and infection in earlier periods. More recently, following the sociological tradition of Louis Wirth, Milgram's concept of Information Overload (1970) suggested a mechanism by which the vulnerable nervous system could be adversely affected in cities. This was that the volume of stimuli, both coming from other people and from the ambient level of signals and messages, could be too much for the vulnerable nervous system to process. This was possibly confirmed by a study of people's speed of walking in the centres of urban communities of varying size: the larger the place, the faster people walked (Bornstein and Bornstein 1976). The authors' interpretation of that study was that in the presence of large numbers of other people, interactions with them have to be reduced, to avoid overloading of the nervous system, though it is clearly open to other interpretations.

Scientifically, though, the evidence remains uncertain. The reason for this is probably that the usual urban–rural dichotomy, which is used in studies, no longer represents a meaningful difference. For one thing, the term 'urban' can mean anything from a town of 20 000 people to a city of nearly 20 million. Most existing data do not distinguish between the various orders of magnitude of urban populations. Cities certainly do contain concentrations of multiply-disadvantaged people, many of whom may have poor mental health along with other unfavourable attributes, but this picture should not be over-generalized.

Epidemiological studies in North Carolina (Blazer *et al.* 1985) showed that major depression was much more common in the urban counties than in the rural ones. But it was concluded there that place of residence now serves largely as a proxy for social processes that affect the risk of developing psychiatric disorder. In other words, that vulnerable people tend to live in certain kinds of places rather than that the conditions of these places cause greater morbidity. It may also be wrong to think of environmental stress as primarily an urban phenomenon. Rural people have always been vulnerable to such stresses as bad weather and social isolation. Social disintegration, insofar as it is an identifiable process, affects many rural areas as well as inner cities, for instance the mountainous areas of Spain and Italy which have been largely abandoned by their former peasant populations (Barquero *et al.* 1982). Though it is widely believed that people in remote areas of the world suffer less emotional distress and fewer psychiatric disorders than are found elsewhere, a survey of two remote mountain villages in Chitral offered no support for this view; the women there were thought to suffer more anxiety and depressive disorders than those in Western societies (Mumford *et al.* 1996).

High density of population is in fact an essential positive aspect of towns and cities. It is needed for social support and to supply a large-enough market for culture and entertainment, and for high-quality services of all kinds, including health services. Otherwise, as Oscar Newman pointed out (1980), there is simply the multiplication of relatively inferior services throughout innumerable suburbs. Most essential services need to be within walking distance of people's homes, and there needs to be a diversity of activities within each neighbourhood (Sherlock 1991). Regrettably, the trends of this century, particularly in North America, have all been in the opposite direction.

There was once a close proximity of work and residence in many places, which had positive aspects. Unfortunately, for many years, planners, architects, and developers have been wedded to the principle of separation of land uses. As a result, shopping centres, housing estates, business parks, industrial zones, and sometimes educational zones have been separated into one-purpose areas, often with great difficulty of access between them. The social results of this have been almost uniformly disastrous. The dispersal and differentiation of human activities in this way makes the control of deviant behaviour more and more difficult. Many well-functioning urban communities have been destroyed through a misconceived pursuit of the goal of low density; but in any case, is there any reason to believe that high density of population is bad for mental health?

CROWDING AND DENSITY

The relationship of crowding and density to each other, and to mental health remains fairly confused. Density is the number of people resident in a geographical unit of area, purely a mathematical ratio. Crowding is the number of people in an area of dwelling. It varies mostly with local socio-economic circumstances, whereas density does not. Therefore, the two may not necessarily correspond at all.

What people regard as 'crowding' is strongly affected by culture, and is largely a psychological experience. In crowding, people feel that their personal space has been invaded (Freedman 1975). Social interaction then becomes involuntary and unpredictable, resulting in high arousal of the central nervous system, and so of the endocrine and other systems. There is also a higher rate of life events, which may have an adverse effect, particularly in precipitating depressive illness. However, the stress associated with this experience of crowding probably derives more from disordered social relationships than from actual lack of physical space. People may actually welcome intense physical crowding at certain times, for instance at a party or in the bar of a pub.

In parts of Asia, very low rates of household space per person do not yet seem to be associated with the kinds of adverse results that would be expected in European or American populations. This applies both to psychiatric disorder and to social pathology. In Hong Kong, I have visited an enormous public housing development, and have been astonished at the high standards of maintenance and cleanliness, and the peaceful general atmosphere. In any similar situation in Britain or North America, such standards would be inconceivable. This better situation seems to result from different social norms on the sharing of space which have been incorporated into that culture, particularly amongst those related by kinship. Individual privacy is felt to be less important than it is in Western populations (Anderson 1972).

In Western societies, cross-sectional analyses of population density, as opposed to crowding, have in fact failed to show any significant relationships with psychiatric morbidity. One likely reason is that the range of densities found in industrialized countries is now fairly narrow. Another reason is that longitudinal studies over time are really needed, rather than cross-sectional ones, and these are extremely difficult to complete. For one thing, very few funding bodies are prepared to commit themselves to supporting studies for the necessary length of time.

COMPREHENSIVE REDEVELOPMENT

'Comprehensive redevelopment' or 'urban renewal' ignored the fact that an established physical environment incorporates a social structure. Gans (1967) pointed out that a physical slum is not necessarily a social slum. In British industrial areas up to the 1950s, the street itself was an extension of the space inside the small houses; a rich pattern of social interaction went on within it, whereas vehicle movements were very few. Since the social structure of an environment is undoubtedly important for the health and well-being of those people who live there, the loss of that structure must be seriously damaging to them. The value of cultural continuity – of 'psychological conservation' – has been completely ignored in most major redevelopments.

Over thirty years ago, Fried (1963) reported a classic study of a working-class Italian community in Boston, which was compulsorily relocated through redevelopment. He found that very high rates of depression resulted from this experience, and that these persisted in many cases for the two years of follow-up. He called this experience 'Grieving for a lost home'. In Britain, no comparable research has ever been carried out, in spite of the enormous scale of urban redevelopment and of new town construction. In Salford, people found their environment radically altered, when they were moved from the familiar 'Coronation Street' to a high-rise block. The clearance of old housing has often been accompanied by grandiose road developments, imposing enormous social costs and bringing no benefits to the usually poor communities through which they pass. Their effects are particularly bad on children, who in many urban situations can no longer walk or cycle anywhere safely. Both children and adolescents find these redeveloped environments unfriendly, with no opportunities for positive social interaction. The result is to encourage vandalism and delinquency.

There may in fact be a special relationship between urban life and the mental health of children. A classic study by Rutter *et al.* (1974) compared ten-year-old children in the Isle of Wight with others in inner London. Both psychiatric morbidity and delayed reading were twice as common in the London children. The correlates of this morbidity – family disturbance, parental deviance, social disadvantage, and school difficulties – were also much more common in London. It seemed that environmental effects were mediated, through families and schools, to the children. Perhaps children are more susceptible than adults to these effects, because they have less control over resources and fewer coping abilities. Nevertheless, the exact nature of this inner city effect on children's psychiatric morbidity has not yet been identified. It was not just poverty.

British studies also provide some evidence that life in high-rise blocks has adverse psychological effects on both young children and their mothers. It is difficult to isolate the factor of height from other characteristics that tend to be associated with it, but there are still a number of strong objective reasons why life in high flats is often stressful (Freeman 1993):

- structural risks, such as Ronan Point (destroyed by a gas explosion)
- the danger of children falling from balconies or windows
- how the play of young children can be safely supervised
- inadequate fire precautions
- the worsening of many social problems through isolation or vandalism
- failure of maintenance of the blocks
- a dehumanized and demoralizing appearance of the environment.

In many places, high-rise blocks were so unsatisfactory – both from the constructional and the social points of view – that they have been demolished. The unfortunate citizens will go on paying off the loans for them until well into the next century.

CRIME AND FEAR OF CRIME

A factor of increasing importance in cities, particularly large cities, is that they contain more crime, and social pathology of all kinds, than less densely populated areas. Therefore, both fear of crime, seen as chronic anxiety disorder, and the consequences of crime, such as Post-traumatic Stress Disorder, must be present at a higher rate in cities. Such an excess must make a significant contribution to mental ill-health there, though this has never been properly investigated. Fear of crime also causes social withdrawal, which undermines supportive social networks.

An important sociological concept here is that of 'incivilities', which can be either physical, social, or mixed. They include environmental degradation such as litter and graffiti, as well as frightening behaviour. These symbols of danger and environmental decline raise anxiety levels and discourage people from using public areas in cities. As a result, the situation becomes even worse. People who have been victims of crime are more upset by incivilities than those who have not been victimized. This relates to the 'broken window' hypothesis: once a single window is broken and not repaired, the remaining windows in an empty building are smashed. Waddington (1996) points out that, 'Crime flourishes in conditions devoid of an authoritative order . . . Maintaining low-level order sustains a virtuous cycle that militates against more serious crime'. Emphasizing the symbolic importance of patrolling police, he adds that 'the best officer represents the authority of the state over public space'.

Rates of crime can be influenced by different architectural policies. The work of Newman in New York introduced the very important concept of 'Defensible Space' (1972). It was shown that changes to public housing can promote the informal control by residents which is the strongest deterrent to crime – the 'Eyes in the Street'. In Britain, Coleman (1985) showed that many of the common architectural features of flats, such as connecting walkways and 'confused' space for which no one was responsible, greatly increased local crime rates. More concern for this aspect of environmental change could therefore be helpful in reducing some aspects of mental ill-health in cities.

On the other hand, the catastrophic loss of economic activity from many Western cities has deprived their communities of access to employment. This has been to such an extent, particularly for young men, that crime has become almost inevitable. Much of it is now associated with the use and sale of drugs. Economic activity now tends to grow on the periphery of cities, which poor people are often unable to reach for work. The demoralization of such communities, the loss of Durkheim's 'moral order of society', is becoming a matter of increasing general concern to Western populations. There is evidence of it in the destructive rioting that occurs from time to time. What is most ominous is the increasing polarization between what J.K. Galbraith has called the 'contented majority' on the one hand and a minority who are increasingly excluded from the normal business of society on the other, and who have been described as the 'underclass'.

CONCLUSION

If adverse environments can have an unfavourable effect on mental health, can mental health be enhanced by improving the environment? The scientific literature contains very little reliable information on this point, mainly because hardly any populations have been studied both before and after such a change. However, a study recently reported by Halpern (1995) at least provides some firm data. He took a random population sample from a peripheral housing estate, which the local Council was proposing to improve comprehensively. This sample was given a questionnaire estimating levels of anxiety and depression, both of which were present to a surprisingly large extent. There was also a generally high level of social distrust and withdrawal. After the improvements, both symptoms were significantly reduced, but anxiety more than depression; social withdrawal was also reduced. Halpern proposed that mental improvement was not just due to the structural changes, but also to the consultation process which preceded them. These data are very encouraging, but need to be replicated.

Reference has already been made to the enormous range of size of today's urban settlements, and with the growing phenomenon in the USA of 'edge cities', apparently unrelated to any central focus, it may be difficult now to decide where a city begins or ends. As yet, we know practically nothing about people who live in the now huge cities of South America, Asia, and Africa. Yet of the world's ten largest cities, four are now in Latin America: Mexico City, São Paulo, Buenos Aires, and Rio de Janeiro. Some of these, such as Mexico City, have already reached sizes never previously known in human history.

Psychiatrists cannot provide a blueprint for society. What they can do, though, is to draw attention to environmental factors which are harmful to mental health, and to try and get this knowledge incorporated into the political process. They could do this task better if there was more systematic study of the topics mentioned here. What is particularly needed is more studies of individual environmental factors, in specific population groups, and in relation to specific psychiatric disorders.

The environment has not only physical aspects, but social, cultural, and aesthetic meanings. These are extremely difficult to investigate, but it should remain our aim to do that to the highest scientific standards. Finally, there is a great need to integrate the existing knowledge of many different disciplines which are relevant to this subject. This is something we could begin to do immediately.

REFERENCES

Anderson, E.N. Jr. (1972). Some Chinese methods of dealing with crowding. *Urban Anthropology*, **1**, 141–50.

Bamrah, J.S., Freeman, H.L., and Goldberg, D.P. (1991). Epidemiology of schizophrenia in Salford, 1974–84: changes in an urban community over ten years. *British Journal of Psychiatry*, **159**, 802–10.

Barquero, J.L.V., Munez, P.E., and Jaurequi, V.M. (1982). The influence of the process of urbanization on the prevalence of neurosis. *Acta Psychiatrica Scandinavica*, **65**, 161–70.

Birtchnell, J., Masters, N., and Deahl, M. (1988). Depression and the physical environment: a study of young married women on a London housing estate. *British Journal of Psychiatry*, **153**, 56–64.

Blazer, D., George, L.K., Landerman, R., *et al.* (1985). Psychiatric disorders: a rural/urban comparison. *Archives of General Psychiatry*, **42**, 651–6.

Bornstein, M.H. and Bornstein, H.G. (1976). The pace of life. *Nature*, **259**, 557–9.

Brown, G.W. (1991). A psychosocial view of depression. In *Community psychiatry: the principles* (ed. D.H. Bennett and H.L. Freeman). Churchill Livingstone, London.

Brown, G.W. and Harris, T. (1978). *Social origins of depression*. Tavistock, London.

Buglass, D. and Duffy, J.C. (1978). The ecological pattern of suicide and parasuicide in Edinburgh. *Social Science and Medicine*, **12**, 241–53.

Castle, D., Scott, K. Wessely, S., *et al.* (1993). Does social deprivation during gestation and early life predispose to later schizophrenia? *Social Psychiatry and Psychiatric Epidemiology*, **28**, 1–4.

Cohen, S., Evans, G.W., Stokols, D., and Krantz, D.S. (1986). *Behavior, Health and Environmental Stress*. Plenum, New York.

Coleman, A. (1985). *Utopia on trial: vision and reality in planned housing*. Hilary Shipman, London.

Dauncey, K., Giggs, J., Baker, K., *et al.* (1993). Schizophrenia in Nottingham: lifelong residential mobility of a cohort. *British Journal of Psychiatry*, **163**, 613–19.

Digby, A. (1985). *Madness, morality and medicine*. Cambridge University Press, Cambridge.

Dubos, R. (1972). *A God within*. Scribners, London.

Durkheim, E. (1897). Le suicide (translated 1952 as *Suicide: a study in sociology*.) Routledge and Kegan Paul, London.

Faris, R.E. and Dunham, H.W. (1939). *Mental disorders in urban areas*. Hafner, Chicago.

Freedman, J.L. (1975). *Crowding and behavior*. W.H. Freeman, San Francisco.

Freeman, H.L. (1993). Mental health and high-rise housing. In *Unhealthy housing: research, remedies and reform* (ed. R. Burridge and D. Ormandy), pp.168–90. Spon, London.

Freeman, H.L. (1994). Schizophrenia and city residence. *British Journal of Psychiatry*, **164** (Supp 23), 39–50.

Freeman, H.L. and Alpert, M. (1986). Prevalence of schizophrenia in an urban population. *British Journal of Psychiatry*, **149**, 603–11.

Fried, M. (1963). Grieving for a lost home. In *The urban condition* (ed. L.J. Duhl). Basic Books, New York.

Gans, H.J. (1967). Planning and city planning for mental health. In *Taming megalopolis* (ed. H.W. Eldredge) Vol 2. Doubleday, New York.

Halpern, D. (1995) *Mental health and the built environment*. Taylor and Francis, London.

Hare, E.H. (1956). Family setting and the urban distribution of schizophrenia. *Journal of Mental Science*, **102**, 753–60.

Leighton, A.H. (1959). *My name is legion*. Basic Books, New York.

Levy, L. and Rowitz, L. (1974). *The ecology of mental disorder*. Behavioral Publications, New York.

Lewis, G., David, A., Andreasson, S., *et al.* (1994). Schizophrenia and city life. *Lancet*, **340**, 137–40.

Milgram, S. (1970). The experience of living in cities. *Science*, **167**, 1461–8.

Mumford, D.B., Nazir, M., Jilani, F., and Baig, I.Y. (1996). Stress and psychiatric disorder in the Hindu Kush: a community survey of mountain villages in Chitral, Pakistan. *British Journal of Psychiatry*, **168**, 299–307.

Newman, O. (1972). *Defensible space*. Macmillan, New York.

Newman, O. (1980). *Community of interest*. Anchor Doubleday, New York.

Park, R.E. and Burgess, E.W. (1925). *The city*. University of Chicago, Chicago.

Rutter, M., Yule, B., Quinton, D., Rowlands, O., Yule, W., and Berger, M. (1974). Attainment and adjustment in two geographical areas. *British Journal of Psychiatry*, **125**, 520–33.

Sainsbury, P. (1955). *Suicide in London*. Chapman and Hall, London

Sherlock, H. (1991). *Cities are good for us*. Transport 2000, London.

Tarnopolsky, A. and Clark, C. (1984). Environmental noise and mental health. In *Mental health and the environment* (ed. H.L. Freeman), pp. 250–70. Churchill Livingstone, London.

Taylor, S. and Chave, S.P.W. (1964). *Mental health and environment*. Longman, London.

Trichopoulos, D., Katsayanni, K., Zavitsanos, X., Tzonai, A., and Dalla-Vorgia, P. (1983). Psychological stress and fatal heart attack: the Athens (1981) earthquake natural experiment. *Lancet*, **1**, 441–4.

Waddington, P. (1996). Finding a real job for Bobby. *The Independent*, 1 March.

Wedmore, K. and Freeman, H.L. (1984). Social pathology and urban overgrowth. In *Mental health and the environment* (ed. H.L. Freeman), pp. 293–326. Churchill Livingstone, London.

Youssef, H.A., Kinsella, A., and Waddington, J.L. (1991). Evidence for geographical variations in the prevalence of schizophrenia in rural Ireland. *Archives of General Pyschiatry*, **48**, 254–8.

7

Meeting of minds: the import of family and society

Anthony W. Clare

Professor Anthony Clare took his medical degree at University College, Dublin and undertook his psychiatric training at St Patrick's Hospital in the same city. He was appointed Psychiatric Registrar at the Maudsley Hospital, London in 1970 and became Senior Registrar three years later. He then moved to the Institute of Psychiatry to undertake research and was appointed Senior Lecturer there in 1980. In 1983, he became Professor and Head of the Department of Psychological Medicine at St Bartholomew's Hospital Medical School; six years later, he returned to his native Dublin to take up his present posts of Clinical Professor of Psychiatry at Trinity College and Medical Director of St Patrick's Hospital. Since 1980, Professor Clare has been a regular broadcaster on radio and television: his two radio series In the Psychiatrist's Chair *and* All in the Mind *have become media classics. He has also published extensively and his book* Psychiatry and Dissent *(1976) has achieved classic status. Professor Clare played a prominent part in Ireland's national debate about the reform of laws on divorce.*

Our ideas about the brain and the mind reflect the science and the politics of the era in which we live. The very cytoarchitectonics of the cerebral cortex are sculpted by inputs from the social environment because socialization shapes the essential human attributes of our species. In practice, most of us, save perhaps the most incorrigibly reductive, conceptualize human life, our own and that of others, in terms of an interaction between the extraordinary organic computer that is the human brain and the equally complex, multifaceted organism that we term society. There is no mental function without brain and social context. The distinguished American psychiatrist, Leon Eisenberg, makes the point with characteristic pungency:

> ... to ask how much of mind is biological and how much social is as meaningless as to ask how much of the area of a rectangle is due to its width and how much to its height or how much of the phenotype is due to genes (nature) and how much to environment (nurture). (Eisenberg 1995*a*)

And yet what goes to form that elusive ideal, the integrated, healthy, mature, and socially functioning adult human being, remains the subject of intense controversy. The current fashion is genetics but then genetics has always been emphasized. The eugenics movement envisaged the infant as essentially genetically constructed and the perfection of the human race as dependent on the weeding out of substandard children – those with bad health, low intelligence, or defective moral character. Leaving aside such determinist views, the scientific evidence supporting the important role played by hereditary factors and organic damage or dysfunction of the brain in the genesis of emotional and behavioural health and illness is now substantial.

Others, most notably behaviourists such as John B. Watson, argued that children are entirely or for the most part the products of their environment. Watson (1928) was particularly uncompromising:

> Treat them as though they were young adults. Dress them, bathe them with care and circumspection. Let your behaviour always be objective and kindly firm. Never hug and kiss them, never sit them on your lap. If you must, kiss them once on the forehead when they say goodnight. Shake hands with them in the morning. Give them a pat on the head if they have made an extraordinary good job of a difficult task.

Behaviourism became the dominant psychological school in the United States during the second quarter of this century when traditional ways of rearing children were repudiated in favour of a new and more scientific approach which the increasing complexity of life seemed to require. According to Bruno Bettelheim, himself a prominent theorist on parent–child relationships (Bettelheim 1987), it has remained America's prevalent psychological doctrine ever since, although others would grant that pre-eminent position to psychoanalysis. Psychoanalytic theory did emphasize certain hereditary aspects, most notably the sexual and aggressive drives, but in general its emphasis on early childhood development, the relationship of the child to its parents, and in particular to its mother has been its real contribution to the theory of personal relationships. Freud himself, Carl Jung, Melanie Klein, Anna Freud, and Donald Winnicott are just some of the psychoanalysts who powerfully argued the central, crucial position of parent–child relationships in the development of the human psyche. John Bowlby's attachment theories were amongst the most important. He argued that it is in our first relationship, usually with our mother, that much of our future well-being is determined. In sharp contrast to both geneticists and behaviourists, Bowlby (1967) believed that the infant needs and seeks affectionate relationships:

> When a baby is born he cannot tell one person from another and indeed can hardly tell person from thing. Yet, by his first birthday he is likely to have become a connoisseur of people. Not only does he come quickly to distinguish familiars from strangers but amongst his familiars he chooses one or more favourites. They are greeted with delight; they are followed when they depart; and they are sought when absent. Their loss causes anxiety and distress; their recovery, relief and a sense of security. On this foundation, it seems, the rest of his emotional life is built – without this foundation there is risk for his future happiness and health.

Bowlby's work, as Robert Karen's masterful recent review makes plain (Karen 1994), was controversial and rejected by most of his peers. Studies carried out in the 1960s and 1970s seemed to indicate that replacement parental figures for children, if they were warm and supportive, meant that the mother and father roles and the attachment process for children were not all-important. Bowlby was savaged by critics who interpreted his emphasis on maternal deprivation as an attack on the mother working outside the home. Developing his theory of attachment Bowlby argued that because we have lost the abundance of secondary attachment figures – aunts, uncles, older siblings, extended family members – the attachment to a parental figure has become more important to children and consequently the loss. Researchers, however, had been slow to look at the link between the attachment and parental loss theory and the effects of separation and divorce on children. Yet thirty years after Bowlby's original arguments concerning the effects of deprivation and disadvantage on children's psychological development, Michael Rutter's painstaking and dispassionate review of the accumulating literature concluded that they 'have been amply confirmed' (Rutter 1981). Rutter emphasized the complex nature of parent–child attachment and went on to argue for the importance of such factors as the multiplicity of stresses, changes in family circumstances, factors in the child and in the children, coping skills and factors outside the home in determining the positive or negative outcome of a child's development. Notwithstanding such elegant dissection, the cruder view of the significance of child–parent relationships summed up in

Philip Larkin's infamous quatrain has tended to dominate thinking in the last two decades of this century:

> They fuck you up, your mum and dad,
> They may not mean to, but they do.
> They fill you with the faults they had
> And add some extra, just for you.

Recent concern about the future of the family has caused a resurgence of research interest in the whole issue of attachment and of the importance of the biological parents to the development of the child. The basic unit of human society is the family, the social grouping composed of a married or unmarried couple and children, including adopted ones. We are all significantly shaped and marked by experience of family – even when, as in the case of the orphaned, it is a shaping and disfiguring by absence. At the heart of the Judaeo-Christian story, of the great tragedies of Greek drama, of the influential speculative theories of Freudian and post-Freudian psychoanalysis, of the major themes and influences of Western thought and art and politics is the family, that most fundamental of our human relationships involving child–parent and sibling relationships.

The stages of the evolution of the modern family, have been elegantly described in Lawrence Stone's monumental work, *The family, sex and marriage in England 1500–1800* (Stone 1977). In the sixteenth century and for at least a millenium before it there was the Open Lineage Family followed by the Restricted Patriarchal Nuclear Family, which began in the early sixteenth century and ran on for several centuries before being replaced in turn by the Closed Domesticated Nuclear Family which began to evolve in the late seventeenth century, predominated in the eighteenth and with some development and wider diffusion has survived well into the nineteenth and the twentieth centuries. This Closed Domesticated Nuclear Family was marked by the rise of what Stone has termed 'Affective Individualism'. It was a family organized around the principle of personal autonomy and bound together by strong affective ties. Husbands and wives personally selected each other rather than obeying parental wishes and their prime motives were now long-term personal affection rather than economic or status advantage for the lineage as a whole. More and more time, energy, love, and money of both parents were devoted to the upbringing of children, whose wills it was no longer thought necessary to crush at an early age. Education loomed larger and larger in parents' minds and among the wealthier members of society the new orientation towards children led to an active desire to limit their numbers in order to improve their life-chances. The home became a more private place. Externally it was closed off from prying neighbours. It was also more closed off from kin, whose role in the choice of marriage partners withered away. After marriage, social ties with relatives were increasingly confined to parents, uncles, and aunts.

It is not well appreciated that the four key features of the modern family – intensified affective bonding of the nuclear core at the expense of neighbours and kin, a strong sense of individual autonomy and the right to personal freedom in the pursuit of happiness, a weakening of the association of sexual pleasure with sin and guilt, and a growing desire for physical privacy – were all well established by 1750 in the key middle and upper sectors of English society. The nineteenth and twentieth centuries merely saw their much wider social diffusion. The development of the behavioural sciences during the past one hundred years, and in particular developmental psychology and psychoanalytic psychotherapy, elevated discussion of the importance of very early childhood in character formation, personal growth, sexual maturation, and the cultivation of a moral sense in adult life.

A century on from Freud, however, there is a considerable ambivalence and unease concerning the state of personal relationships. Marriage is perceived to be in crisis. The four key features of modern marriage are beginning to be perceived as hopelessly idealistic and ultimately destructive. Divorce rates are high – in the US the figure is about 55 per 100 marriages, in Canada 38, in the UK in excess of 50. Children living with one parent or with one biological and one step-parent are increasing in number as are so-called 'melded' sibships joining children from several families. Revelations concerning childhood sexual abuse shake the benevolent view of the family as a crucible of affection and psychological health. The size, composition, structure, and living arrangements and circumstances of families in North America and Europe have undergone changes so profound as to raise the question: is the family as we have known it becoming outmoded and obsolete (Eisenberg 1995*b*)? It is in this context that certain aspects of the relationship between marital and family relationships and health and social functioning merit reconsideration.

The number of children under 16 in Great Britain whose parents have divorced increased by one-sixth between 1977 and 1992. The increase has been especially marked for children under five. One in four children will experience parental divorce by age 16 and

13% of children will live in a stepfamily formed by remarriage or cohabitation. In the United States, 50% of all children are predicted to see their parents divorce (McAllister *et al.* 1995). A considerable body of research testifies to the short-term and the long-lasting impact of marital breakdown and divorce on children. Such research also demonstrates that this impact often extends into adult life with consequences for health, family life, educational performance, and occupational status. However, more recent reviews of research in this area do highlight methodological complexities and the importance of accounting for pre-divorce effects in measuring post-divorce outcomes. Demo and Adcock (1991) put the issue squarely:

> It is simplistic and inaccurate to think of divorce as having uniform consequences for children. The consequences of divorce vary along different dimensions of well-being, characteristics of children (e.g. pre-divorce adjustment, age at time of disruption) and characteristics of families (e.g. socioeconomic history, pre- and post- divorce levels of conflict, parent-child relationships and maternal employment). Most of the evidence reviewed . . . suggests that some sociodemographic characteristics of children such as race and gender are not as important as characteristics of families in mediating effects of divorce.

One of the earliest comprehensive reviews of the impact of divorce on children in Britain concluded that 'marital separation is a process with profound consequences for children' and the authors noted that the most common reactions in affected children included anger directed at one or both parents, sadness, and depression (Richards and Dyson 1982). In younger children, clinging to parents and regressive reactions like bedwetting were frequently seen while older children reportedly withdrew somewhat from the home and sought relationships elsewhere. Butler and Golding's analysis of the MRC National Survey of Health and Development 1970 birth cohort confirmed this tendency towards 'regressive' reactions. Children whose parents divorce before they are five years were 50–100% more likely to bedwet, soil, or throw temper tantrums. Similar evidence emerged from the 1946 birth cohort study where bedwetting amongst children who had experienced parental divorce before the age of six was more than twice the rate of those who parents had not divorced (Douglas 1973). In addition, children of divorced parents were 50% more likely to be admitted to hospital (Butler and Golding 1986) or to suffer accidental injury or poisoning (Dawson 1991).

Different options appear available to children of different ages in terms of responding to parental divorce (Hetherington *et al.*

1992). Adolescent children in divorced lone mother families and in stepfamilies formed through remarriage consistently scored less well on indices of behaviour, competence, and education than comparable children, whose parents remained married. Fewer sex differences are found in studies of adolescent children than in studies of younger children, with adolescent girls as likely to 'act out' as boys. Nor are the effects temporary or short-term. In the 1946 cohort study, subjects were interviewed, tested, or examined throughout childhood, at 21 years, 26 years, 31 years and, in 1982, at 36 years. The findings reveal a disturbing fact: the experience of divorce as a child can have adverse effects in terms of health, behaviour, and economic status some thirty years later. There is evidence that those whose parents divorce when they are less than five years old are particularly vulnerable (Wadsworth 1984). Children of divorce fare worse—and do so persistently in terms of educational attainment and behavioural measures too. Analysing data from the first four sweeps of the 1958 National Child Development Study, whose respondents were aged 7, 11, 16, and 23 respectively, Elliott and Richards (1991) noted that children whose parents divorced between sweeps of the study were already scoring lower on a range of measures in the sweep *prior* to their parents' divorce than were children whose parents remained married. These authors suggested that factors apart from divorce itself may be at work including parental conflict, social class, and children's age at the time of their parents' break-up. As Dominian and his colleagues point out in a solid and dispassionate review of this entire area, the fact that lower attainment was already evident in children before divorce occurred underlines the need to understand marital breakdown as a process (McAllister *et al.* 1995).

There are health consequences too of marital breakdown and family disruption. Children of divorced parents appear more susceptible to developing psychiatric ill-health. Men whose parents had divorced before they were five years old were three times as likely to suffer from psychiatric illness as those from intact families (6.3% compared with 1.7%). Those aged 16–26 at the time of their parents' divorce experience an even greater incidence of psychiatric illness; a level of 9.8% was found in disrupted families – five times the rate of those from intact families. Psychiatric illness is twice as common amongst women with divorced parents as for those from intact families – 3.5% compared with 1.5% (Wadsworth 1984). Even at the age of 36, women from disrupted homes were reporting a higher number of psychiatric symptoms (Kuh and Maclean 1990). There is widespread evidence of increased behavioural problems and delinquency among both boys and girls whose parents have divorced. Analysis of data from the second National Health Interview Survey on Child Health in the USA has shown that children living with single mothers or with mothers and stepfathers were two to three times more likely to have to repeat a grade at school, to have been suspended or expelled from their school and three times as likely to be in need of treatment for emotional or behavioural problems (Dawson 1991). Similarly, children from a background of divorce or separation scored higher on measures of antisocial behaviour, anxiety, depression, inattention, hyperactivity, dependency, and fearfulness (Dawson, 1991; Guidubaldi *et al.* 1987 *a,b*). Adjusting for social class and ethnic factors does reduce the significance of these findings; nevertheless even when income and ethnicity are taken into account children from single parent families fare worse. As in other studies, the boys were more severely affected. Delinquent behaviour is particularly increased for those whose parents divorced before they were five years old. Nearly 30% of such boys had shown delinquent behaviour by the age of 21 compared with 14% of boys from intact families. Whilst girls are much less likely to be delinquent, 7.9% of those whose parents had divorced showed delinquent behaviour by the age of 21 compared with only 1.6% of those from intact families (Wadsworth 1984). Demo and Adcock (1991) report that both girls and boys from backgrounds of divorce in the United States also have elevated rates of delinquency.

There is a remarkable degree of unanimity concerning these adverse effects of parental divorce on children's educational development. These effects can occur at any age. For example Allison and Furstenberg (1989) found consistent differences in their analysis of the second round of the National Survey of Children between six-year-old children from intact and disrupted families on social, behavioural, and academic measures.

More specifically, children of divorced parents were more likely to be placed in special reading classes, to be doing less well academically and socially, and were more likely to have been referred to an educational psychologist. Following parental divorce the chances of attending university are considerably reduced (Maclean and Wadsworth 1988). For those from a non-manual background, the chances of a university education are twice as great if the individual comes from an intact family as compared to one affected by divorce. A comparable differential occurs for those from a manual working background and for 'A' level results. Men who have experienced parental divorce suffer with respect to their earning capacity, with twice the proportion expected found in the lowest income bracket.

The 1958 cohort confirms these trends. Those whose parents had divorced when they were aged between 7 and 16 were more likely to leave school without qualifications and less likely to go to university than children whose parents were still married (Elliott and Richards, 1991). A Finnish study based on all pupils attending a large secondary school (Aro and Palosaari 1992) found that less than half as many children of divorced parents went on to university compared with children of intact marriages:

> These findings are of great importance because those sceptical or unaware of the studies of the effects of divorce on children claim that the observed differences are the result of economic factors. Accordingly they argue that children suffer because the standard of their living falls. While it is undoubtedly true that the fall of economic standards has attendant short comings, for example, changes of housing or moving school, it must be recognised that the evidence from research suggests that other factors are in play. Emotional disturbance and stress are particularly notable in the critical early years of childhood.

In discussing the importance of the family, violence – one of the central preoccupations of late twentieth society – should be considered. In a recent review entitled '*Causes of violence*', the distinguished American psychologist John Monahan (1993) identified amongst the tangled roots biological factors including testosterone, lead, head injury, and poor nutrition, and sociological factors including crime, poverty, unemployment, and youth. He then declared:

> Finally come the psychological causes. If research on violence were like stock on Wall Street, then where I would put my money right now is on psychology. By this I most emphatically do not mean mental disorder. The best epidemiological evidence indicates that major mental disorder accounts for at most three per cent of the violence in American society. What I mean instead are the developmental processes that we all go through, most of us more or less successfully but some of us with great difficulty. I mean particularly the family – the filter through which most of the sociological factors, such as parents being unemployed, and many of the biological factors, like poor nutrition, seem to have their effect on a child growing up.

So do children need two parents? Notwithstanding the fact that many children raised by single parents develop well and live lives of value and self-respect while many reared within the conventional two-parent nuclear family do poorly, the answer for once to a complicated socio-behavioural question is clear. Children do need two parents. Analysis of six nationally representative data sets in the United States by McLanahan and Sandefur emphasizes this fact (Eisenberg 1995*b*). Children growing up in single-parent households, whether the parents were never married, separated, or divorced, have twice the risk of dropping out of high school, of being out of work, and of becoming teenage mothers themselves. The split family loses about 40% of its income. Children have much less time with their fathers – one-third see them once a week, another third not at all. The children are twice as likely to have to change residence and school, losing friends in the process. Remarriage being common, many children must adapt to step-parents and step-siblings. One-third of second marriages in the US in turn end in divorce.

Follow-up studies of the children of divorced and separated parents make worrying reading too. One of the earliest and most controversial studies to look at more long-term effects, the California-based research effort by Wallerstein and Blakeslee, recently reported on a ten-year follow-up of children of divorce, many of these children having reached their twenties (Wallerstein and Blakeslee 1989). Many of these young men and women appeared to be profoundly troubled, drifting, and underachieving. Almost all regarded issues of love, commitment, and emotional interdependence with wariness and anxiety. There has been much criticism of this study because it involved an unmatched sample; yet its findings that those who experience the divorce of their parents are less able to form stable intimate relationships with spouses is itself supported by a number of British studies (Maclean and Wadsworth 1988; Kiernan 1986). Women from a background of divorce tend to marry earlier than those from intact families in addition to being more likely to divorce. Kuh and Maclean (1990) in their analysis of women at age 36 have found that 16.3% of children from intact homes had divorced compared with 23% coming from backgrounds of divorce and separation.

In the face of such evidence what social policies can take shape that might support the family and promote child health? First, the role of marriage and the family in the genesis of health and illness must be recognized by political planners as much as by social theorists: this is by no means the case. As Dominian and his colleagues recently pointed out, in over one hundred pages of analysis and plans for action the British Government's highly publicized 1991 Green Paper entitled '*The Health of the Nation*' made not a single mention of the impact of fragmenting personal relationships on physical and mental health (McAllister *et al.* 1995). Yet consider suicide, a particular target for Government action. Platt and his colleagues in a comparative study in Edinburgh and Oxford reported that the relative risk of suicide was around five times as great for divorced as for married men and three times as great for divorced as for married women (Platt *et al.* 1988). Over thirty years ago, Neil Kessel showed in a study of self-poisoning of 68 married men and 147 married women that marital disharmony was a major precipitating factor for 68% of the men and 60% of the women (Kessel 1965).

A twin approach involving the formulation of economic and legal policies to promote family formation and stability and education for responsible parenting would seem sensible. Amongst the suggested elements of such an approach are the better preparation of couples intending to have children, a contract between couples proposing to have children, legislation emphasizing the centrality of children in any separation/divorce proceedings, legislation to ensure that children will continue to be parented by their natural parents after separation/divorce save for special circumstances, mandatory counselling/mediation to consider the future of the children after separation/divorce. Drawing on the work carried out by researchers such as Roslyn Corney and myself in the 1980s, which showed that most people with marital difficulties do not go to marriage guidance counsellors but go to their GPs (Corney *et al.* 1984), it has been argued that what is particularly needed in terms of health service response is appropriately trained professionals within the primary health care team to whom people in trouble, whether as spouses or parents or both, can turn for guidance and assistance.

None of these suggested remedies will alter the situation radically. Even together, it is unlikely that the crisis in family life and stability will be averted; but they do serve to concentrate our thinking and focus our attention on the plight of the many young people whose early years are blighted by marital disharmony and fragmentation. Governments have become preoccupied by the financial cost of supporting children whose parents have divorced and where the absent parent (usually the father) no longer supports the family financially. In Britain this led to the setting up of the Child Support Agency. The evidence concerning the impact of marital separation and divorce on the physical and the psychological health of children is of such an order as to demand not piecemeal responses of this kind but a much more comprehensive and co-ordinated long-term strategy.

Let me end by quoting John Monahan (1993) once again. He suggests a modest idea to combat violence, an idea deriving from research on child-rearing that finds parental supervision so important in preventing crime and violence:

> Taking our cue from studies like this, we could offer to a random group of parents whose children are enrolled in federal child-care programs an intensive, long-term, state-of-the-art education and training program in how to monitor their children's behavior effectively, how to recognise potentially serious misbehavior when it occurs, and how to discipline their children consistently but fairly in response to misbehavior. If this worked – if compared to a control group, the children whose parents received the program had lower levels of aggression and other social problems – we should gradually expand the program, rigorously evaluating its effects each step of the way. If it did not work, we should go back to the drawing board, roll up our sleeves and try something different. A dozen ideas like this – none of them panaceas – could be derived from research on children and families and tried simultaneously in different parts of the country. If even a few of them worked, we would have taken a giant leap forward in violence prevention.

If even a few of them worked, we would have taken a leap forward in the prevention of many forms of individual and social pathology.

REFERENCES

Allison, P.D. and Furstenberg, F.F. (1989). How marital dissolution affects children. *Developmental Psychology*, **25**, 540.

Aro, H.M. and Palosaari, U.K. (1992). Parental divorce, adolescence and the transition to young adulthood: a follow-up study. *American Journal of Orthopsychiatry*, **62**, 412–28.

Bettelheim, B. (1987). *A good enough parent*. Thames and Hudson, London.

Bowlby, J. (1967). Foreword. In *Infancy in Uganda: infant care and the growth of love* (ed. M.D.S. Ainsworth). Johns Hopkins, Baltimore.

Butler, N.R. and Golding, J. (1986). *From birth to five: a study of the health and behaviour of Britain's five year olds*. Pergamon Press, Oxford.

Corney, R.H., Cooper, A., and Clare, A.W. (1984). Seeking help for marital problems: the role of the general practitioner. *Journal of the Royal College of General Practitioners*, August, 421–33.

Dawson, D.A. (1991). *Family structure and children's health: United States 1988*. Series 10:178. Vital and Health Statistics Public Health Service, Maryland.

Demo, D.H. and Adcock, A.C. (1991). The impact of divorce on children. In *Contemporary families: looking forward, looking back* (ed. A. Booth). National Council for Family Relations, Minneapolis, Minnesota.

Douglas, J.W.B. (1973). Early disturbing events and enuresis. In *Bladder control and enuresis* (ed. I. Kolvin, R.C. MacKeith, and S.R. Miedow), pp.109–17. Spastics International Medical Publishers, London.

Eisenberg, L. (1995*a*). The social construction of the human brain. *American Journal of Psychiatry*, **152**, 1563–75.

Eisenberg, L. (1995*b*). *Is the family obsolete?* Unpublished paper. Harvard, Boston.

Elliott, B.J. and Richards, M.P.M. (1991). Children and divorce: educational performance and behaviour before and after parental separation. *International Journal of Law and the Family*, **5**, 258.

Guidubaldi, J., Perry, J., and Nastasi, B.K. (1987*a*). Assessment and intervention for children of divorce. In *Advances in family intervention, assessment and theory*, pp.36–9. JAI Press, Greenwich, Connecticut.

Guidubaldi, J., Perry, J., and Nastasi, B.K. (1987*b*). Growing up in a divorced family. In *Annual review of applied social psychology* (ed. S. Oskamp), pp.202–37. Sage Publications, Beverly Hills.

Hetherington, E.M., Clingempeel, W.G., Anderson, E.R., Deal, J.E., Hagan, M.S., Hollier, E.A., *et al.* (1992). *Coping with marital transitions.* Monographs of the Society for Research in Child Development, Series No. 227, Vol. **57**, No.2–3. University of Chicago Press.

Karen, R. (1994). *Becoming attached.* Warner Books, New York.

Kessel, N. (1965). Self-poisoning – Part 1. *British Medical Journal*, **2**, 1265.

Kiernan, K.E. (1986) Teenage marriage and marital breakdown: a longitudinal study. *Population Studies*, **40**, 35.

Kuh, D. and Maclean, M. (1990). Women's childhood experience of parental separation and their subsequent health and socio-economic status in adulthood. *Journal of Biosocial Science*, **22**, 121.

McAllister, F., Dominian, J., Mansfield, P., Breen, S., Morrod, D., and Percival, J. (1995). *Marital breakdown and the health of the nation* (2nd edn). One Plus One, The Marriage and Partnership Research Charity, London.

Maclean, M. and Wadsworth, M.E.J. (1988). The interests of children after parental divorce: a long-term perspective. *International Journal of Law and the Family*, **2**, 155.

Monahan, J. (1993). Causes of violence. In *Drugs and violence in America* (ed. United States Sentencing Commission), pp. 77–85. US Government Printing Office, Washington DC.

Platt, S., Hawton, K., Kreitman, N., Fagg, J., and Foster, J. (1988). Recent clinical and epidemiological trends in parasuicide in Edinburgh and Oxford: a tale of two cities. *Psychological Medicine*, **18**, 405.

Richards, M.P.M. and Dyson, M. (1982). *Separation, divorce and the development of children: a review.* Department of Health and Social Security, London.

Rutter, M. (1981). *Maternal deprivation reassessed* (2nd edn). Penguin, Middlesex.

Stone, L. (1977). *The Family, sex, and marriage in England 1500–1800.* Weidenfeld and Nicolson, London.

Wadsworth, M.E.J. (1984). Early stress and associations with adult health

behaviour and parenting. *Stress and disability in childhood* (ed. N.R. Butler and B.D. Corner), pp.100–4. John Wright and Sons, Bristol.
Wallerstein, J.S. and Blakeslee, S. (1989). *Second chances: men, women and children a decade after divorce*. Bantam Press, London.
Watson, J.B. (1928). *Psychological care of infant and child*, pp.81–82. Norton, New York.

8

Culture, nature, environment: steps to an ecology of life

Tim Ingold

Professor Tim Ingold has been Max Gluckman Professor of Social Anthropology at the University of Manchester since 1995. He took his first degree, a double First in Natural Sciences and Social Anthropology, at the University of Cambridge and stayed on there to undertake research for his PhD, which he was awarded in 1976. He joined the Department of Social Anthropology at Manchester in 1974 and has served it ever since, as Lecturer, Senior Lecturer, Professor, and Head of Department. He has carried out ethnographic fieldwork among Saami (Lapp) people in northern Finland, and has written extensively on hunter-gatherer and pastoral societies, evolutionary theory, and human ecology. His current research deals with issues of environmental perception and the anthropology of technology. Professor Ingold's major works include Evolution and Social Life (*1986*) *and* The Appropriation of Nature (*1986*).

As a social anthropologist with ethnographic interests in the northern circumpolar regions, I begin with an observation drawn from my own field experience of mustering reindeer in Finnish Lapland. When pursuing reindeer, there often comes a critical point when a particular animal becomes immediately aware of your presence. It then does a strange thing. Instead of running away it stands stock still, turns its head, and stares at you squarely in the face. Biologists have explained this behaviour as an adaptation to predation by wolves. When the reindeer stops, the pursuing wolf stops too, both of them getting their breath back for the final, decisive phase of the episode when the deer turns to flight and the wolf rushes to overtake it. Since it is the deer that takes the initiative in breaking the stalemate, it has a slight head start, and indeed a healthy adult deer can generally outrun a wolf (Mech 1970). But the deer's tactic, that gives it such an advantage against wolves, renders it peculiarly vulnerable on encountering human

Fig. 8.1 The animal offers itself to the hunter. The moment of encounter between an Inuit hunter and a caribou, drawn by Davidialuk, an Inuit artist from Povungnituk in northern Quebec. (Source: Saladin d'Anglure 1979.)

hunters equipped with projectile weapons or even firearms. When the animal turns to face the hunter, it provides the latter with a perfect opportunity to take aim and shoot. For wolves, deer are easy to find, since they travel with the herd, but hard to kill; for humans, to the contrary, deer may be hard to find, but once you have established contact, they are rather easy to kill (Ingold 1980).

The Cree people, native hunters of northeastern Canada, have a different explanation for why reindeer – or caribou as they are called in North America – are so easy to kill. They say that the animal offers itself up, quite intentionally and in a spirit of goodwill or even love towards the hunter. The bodily substance of the caribou is not taken, it is received. It is at the moment of encounter, when the animal stands its ground and looks the hunter in the eye, that the offering is made (Fig. 8.1). As with many other hunting people around the world, the Cree draw a parallel between the pursuit of animals and the seduction of young women, and liken killing to sexual intercourse. In this light, killing appears not as a termination of life but as an act that is critical to its regeneration.[1]

SCIENCE AND INDIGENOUS KNOWLEDGE

Here, then, are two accounts – one from biological science, the other from indigenous people – of what happens when humans encounter reindeer or caribou. How are we to understand the relation between them? Wildlife biologists are liable to react with a mixture of cynicism and incredulity to native stories about animals presenting themselves of their own accord. The cynical view would be that such stories provide a very handy way of dodging the ethical issues surrounding hunting and killing that cause such anxiety for many people in Western societies. For hunters, it is most convenient to be able to transfer responsibility for the death of animals onto the animals themselves. What the Western scientist finds hard to believe is that anybody should be taken in by patently fanciful excuses of this kind. Caribou are being tracked down and killed, for heaven's sake! Could any intelligent person seriously think that animals actually offer themselves to hunters as recounted in the stories of the Cree? Are the folk who tell these stories mad, lost in a fog of irrational superstition, incapable of distinguishing fact from fantasy, talking in allegories, or simply having us on? Whatever the answer may be, science insists that stories are stories, and as such have no purchase on what really goes on in the natural world. Who, after all, would be so foolish as to consult 'Little Red Riding Hood' for a scientific account of the morphology and behaviour of the wolf?

Anthropologists, by temperament and training, are inclined to be more sympathetic towards indigenous accounts. On being told that the success of hunting depends upon the bestowal of favour by animals, the anthropologist's first concern is not to judge the truth of the proposition but to understand what it means. There is a well-tried expository device in anthropology, amounting to a willing suspension of disbelief, that makes it possible to get on with this job of understanding unhampered by worries about whether there is

[1]For detailed ethnographic description of Cree attitudes to animals, see Feit 1973; Tanner 1979; Scott 1989; and Brightman 1993. For further discussion of the idea of animals offering themselves to human hunters, see Ingold 1994.

any foundation in reality for what the people have to tell us. Thus it can readily be shown that the idea of animals offering themselves to hunters, however bizarre it might seem from the viewpoint of Western science, makes perfectly good sense if we start from the assumption (as the Cree evidently do) that the entire world – and not just the human world – is saturated with powers of agency and intentionality. In Cree cosmology, the anthropologist concludes, relations with animals are modelled on those that obtain within the human community, such that hunting is conceived as a moment in an ongoing interpersonal dialogue (Tanner 1979; cf. Gudeman 1986). This is not to say that the biological explanation of the stand-off between hunter and caribou at the point of encounter, as part of an innate response mechanism designed to combat predation by wolves, is without interest. However for anthropologists, explaining the behaviour of caribou is none of their business; their concern is to show how hunters' direct experience of encounters with animals is given form and meaning within those received patterns of interconnected images and propositions that, in anthropological parlance, go by the name of 'culture'.

Although from the above, the perspectives of the wildlife biologist and the cultural anthropologist might seem incompatible, they are in fact perfectly complementary, and indeed disclose a common, albeit practically unattainable, point of observation.[2] Whereas the biologist claims to study organic nature 'as it really is', the anthropologist studies the diverse ways in which the constituents of the natural world figure in the imagined, or so-called 'cognized' worlds of cultural subjects. There are any number of ways of marking this distinction, but of these the most notorious, at least in anthropological literature, is that between so-called 'etic' and 'emic' accounts. Derived from the contrast in linguistics between phonetics and phonemics, the former purports to offer a wholly neutral, value-free description of the physical world, while the latter spells out the specific cultural meanings that people place upon it.

Two points follow from this distinction. First, the suggestion that human beings inhabit discursive worlds of culturally constructed significance implies that they have already taken a step out of the world of nature within which the lives of all other creatures are confined. The Cree hunter, it is supposed, narrates and interprets his experiences of encounters with animals in terms of a system of cosmological beliefs, the caribou does not. Secondly, to perceive this

[2] I return in the concluding section to the reasons why this point of observation cannot, in practice, be attained (see p. 178).

system *as* a cosmology requires that we observers take a further step, out of the worlds of culture in which the lives of all other *humans* are said to be confined. What the anthropologist calls a cosmology is, for the people being studied, a lifeworld. Only from a point of observation beyond culture is it possible to regard the Cree understanding of the relation between hunters and caribou as but one possible construction, or 'modelling', of an independently given reality. By the same token, only from such a vantage point is it possible to apprehend the given reality for what it is, independently of any kind of cultural bias.

It should now be clear why natural science and cultural anthropology converge on a common vertex. The anthropological claim of perceptual relativism – that people from different cultural backgrounds perceive reality in different ways since they process the same data of experience in terms of alternative frameworks of belief or representational schemata – does not undermine but actually reinforces the claim of natural science to deliver an authoritative account of how nature really works. Both claims are founded upon a double disengagement of the observer from the world. The first sets up a division between humanity and nature; the second establishes a division, within humanity, between 'native' or 'indigenous' people who live in cultures, and 'enlightened' Westerners who don't. Both claims, too, are underwritten by a commitment that lies at the heart of Western thought and science, to the extent of being its defining feature. This is the commitment to the supremacy of abstract or

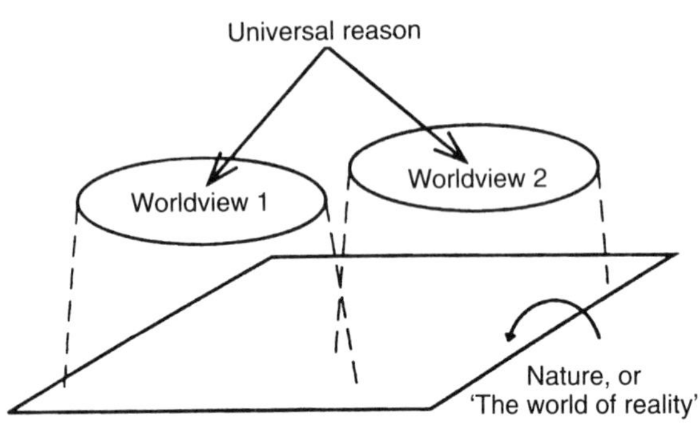

Fig. 8.2 The sovereign perspective of abstract or universal reason, which treats the lifeworlds of people of different cultures as alternative constructions, cosmologies, or 'world-views', superimposed upon the 'real' reality of nature. From this perspective, anthropology embarks on the comparative study of cultural world-views, while science investigates the workings of nature.

universal reason. If it is by the capacity to reason that humanity, in this Western discourse, is distinguished from nature, then it is by the fullest development of this capacity that modern science distinguishes itself from the knowledge practices of people in 'other cultures' whose thought is supposed to remain somewhat bound by the constraints and conventions of tradition. In effect, the sovereign perspective of abstract reason is a compound of two dichotomies: between humanity and nature; and between modernity and tradition. The result is not unlike that produced by perspective painting, in which a scene is depicted from a point of view which itself is given independently of that of the spectator who contemplates the finished work. Likewise abstract reason can consider diverse world-views each of which is a specific construction of an external reality (Fig. 8.2). The anthropologist, surveying the tapestry of human cultural variation, is like the visitor to the art gallery – a 'viewer of views'. Perhaps it is no accident that both perspective painting and anthropology are products of the same trajectory of Western thought (Ingold 1993*a*).

MIND AND NATURE: GREGORY BATESON AND CLAUDE LÉVI-STRAUSS

Let me now introduce the terms of my title. I have shown that the possibility of an objective account of a natural phenomenon such as the behaviour of caribou, as well as the recognition of an indigenous account, such as that of the Cree, as fitting within a particular culture-specific cosmology, depend on a two-step movement of disengagement that cuts out first nature, then culture, as discrete objects of attention. Whereas the scientific account is attributed to disinterested observation and rational analysis, the indigenous account is put down to the accommodation of subjective experience within 'beliefs' of questionable rationality. I wish now to retrace these two steps in the reverse direction. Only by doing so can we overturn the ranking, implicit in what has been said up to now, of scientific over indigenous accounts. I believe that it is necessary to take these steps, to descend from the imaginary heights of abstract reason and resituate ourselves in an active and ongoing engagement with our environments, if we are to arrive at an ecology that is capable of recovering the reality of the life process itself. My aim is to replace the stale dichotomy of nature and culture with the dynamic synergy of organism and environment, in order to regain a genuine ecology of life. This ecology, however, will look very different from what has become familiar to us from scientific textbooks, for it comprises a kind of knowledge that is fundamentally resistant to transmission in an authorized textual form, independently of the contexts of its instantiation in the world.

My subtitle, 'steps to an ecology of life', is borrowed from the work of Gregory Bateson (1973). I have, however, substituted the word 'life' for the word 'mind' as it appears in the title of Bateson's famous collection of essays. This substitution is deliberate. Bateson was a great dismantler of oppositions – between reason and emotion, inner and outer, mind and body. Yet curiously, he seemed unable to shake off the most fundamental opposition of all, between form and substance. His objection to mainstream natural science lay in its reduction of 'real' reality to pure substance, thus relegating form to the illusory or epiphenomenal world of appearances. This he saw as the inevitable consequence of the false separation of mind and nature. Bateson thought that mind should be seen as immanent in the whole system of organism–environment relations in which we humans are necessarily enmeshed, rather than confined within our individual bodies as against a world of nature 'out there'. As he declared, in a lecture delivered in 1970,[3] 'the mental world – the mind – the world of information processing – is not limited by the skin' (Bateson 1973). Yet the ecosystem, taken in its totality, was nevertheless envisaged as two-faced. One face presents a field of matter and energy, the other presents a field of pattern and information; the first is all substance without form, the second is all form detached from substance. Bateson likened the contrast to one which Carl Jung, in his *Seven Sermons to the Dead*, had drawn between the two worlds of the 'pleroma' and the 'creatura'. In the former there are forces and impacts but no differences; in the latter there are only differences, and it is these differences that have effects (Bateson 1973). Corresponding to this duality Bateson recognized two ecologies: an ecology of material and energy exchanges, and an ecology of ideas. And it was this second ecology that he called the 'ecology of mind'.

To bring out the full significance of Bateson's position, it is instructive to set it alongside that of another giant of twentieth century anthropology, Claude Lévi-Strauss. In a lecture on 'structuralism and ecology', delivered two years after Bateson's lecture, Lévi-Strauss likewise set out to demolish the classical dichotomy between mind

[3]The occasion was the Nineteenth Annual Korzybski Memorial Lecture, presented in New York in January 1970, and the text was originally published in the *General Semantics Bulletin*, 1970, Volume 37.

Fig. 8.3 'Day and night' (1938), a woodcut by the Dutch artist M. C. Escher, aptly illustrates, in visual form, the way in which the mind – according to Lévi-Strauss – works upon the data of perception. Drawing upon a selection of recognizable and familiar features of the environment, such as houses, fields, a river, flying swans, the mind casts them into a symmetrical structure of oppositions and contrasts: day/night, left/right, city/country, water/land.

and nature.[4] Although neither figure made any reference to the other's work, there are some superficial resemblances between their respective arguments. For Lévi-Strauss, too, the mind is a processor of information, and information consists in patterns of significant difference. Unlike Bateson however, Lévi-Strauss anchors the mind very firmly in the workings of the human brain. Fastening in a more or less arbitrary fashion upon certain elements or distinctive features that are presented to it in the surrounding environment, the mind acts rather like a kaleidoscope, casting them into patterns whose oppositions and symmetries reflect underlying universals of human cognition (Fig. 8.3). It is by these interior patterns that the mind possesses knowledge of the world outside. If, in the final analysis, the distinction between mind and nature is dissolved, it is because the neurological mechanisms that underwrite the mind's apprehension of the world are part of the very world that is apprehended. This world,

[4]This was the Gildersleeve Lecture, delivered at Barnard College, New York, in March 1972. The text first appeared in *Barnard Alumnae*, Spring 1972, and is reprinted as Lévi-Strauss (1974).

according to Lévi-Strauss, is structured through and through, from the lowest level of atoms and molecules, through the intermediate levels of sensory perception, to the highest levels of intellectual functioning. 'When the mind processes the empirical data which it receives previously processed by the sense organs', Lévi-Strauss concluded, 'it goes on working out structurally what at the outset was already structural. It can only do so in as much as the mind, the body to which the mind belongs, and the things which body and mind perceive, are part and parcel of one and the same reality' (1974).

In all these respects, Bateson's position could not have been more different. For Lévi-Strauss ecology meant 'the world outside', mind meant 'the brain'; for Bateson both mind and ecology were situated in the *relations between* the brain and the surrounding environment (Fig. 8.4). For Lévi-Strauss, the perceiver could only have knowledge

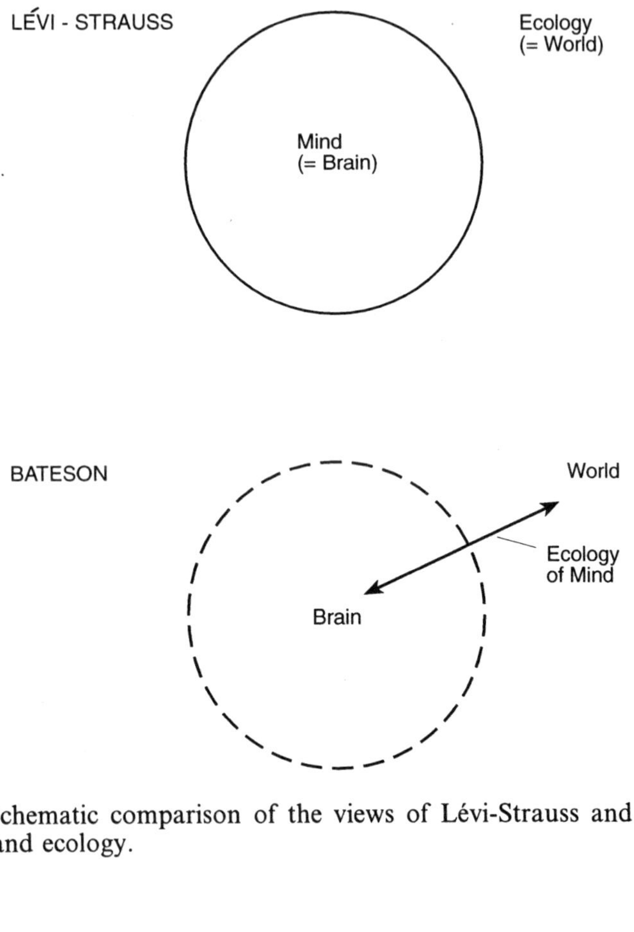

Fig. 8.4 Schematic comparison of the views of Lévi-Strauss and Bateson on mind and ecology.

of the world by virtue of a passage of information across the boundary between outside and inside, involving successive steps of encoding and decoding by the sense organs and the brain, and resulting in an inner mental representation. For Bateson the idea of such a boundary was absurd, a point he illustrated with the example of the blind man's cane (1973). Do we draw a boundary around his head, at the handle of the cane, at its tip, or halfway down the pavement? If we ask where the mind is, the answer would not be 'in the head rather than out there in the world'. It would be more appropriate to envisage mind as extending outwards into the environment along multiple sensory pathways of which the cane, in the hands of the blind man, is just one. Thus while Bateson shared with Lévi-Strauss the notion of mind as a processor of information, he did not regard processing as a step-by-step refinement or repackaging of sensory data already received, but rather as the unfolding of the whole system of relations constituted by the multi-sensory involvement of the perceiver in his or her environment.

To continue with the example of the blind man, it is as though his processing of information were tantamount to his own movement – that is, to his own processing through the world. The point about movement is critical. For Lévi-Strauss, both the mind and the world remain fixed and immutable, while information passes across the interface between them. By contrast, in Bateson's account information only exists thanks to the movement of the perceiver relative to his or her surroundings. Bateson constantly emphasized that stable features of the world remain imperceptible unless we move in relation to them: if the blind man picks up surface features of the road ahead by sweeping his cane from side to side, people with normal vision do the same with their eyes. Through this scanning movement we *draw* distinctions, in the sense not of representing them graphically, but of 'pulling them out'.[5] Whereas Lévi-Strauss often writes as though the world were

[5]Bateson developed this idea in his last work, *Mind and Nature* (1980). There are striking parallels here with ideas developed by two other major writers on the philosophy and psychology of perception, Maurice Merleau-Ponty and James Gibson. Merleau-Ponty, in his essay on *Eye and mind* (1964), insisted on the 'intertwining of vision and [eye] movement', neither of which could occur without the other. Gibson, for his part, placed movement at the heart of his ecological theory of visual perception, regarded as 'a process of information pickup involving the exploratory activity of looking around, getting around, and looking at things' (1979). All three authors stressed the primacy of the perceiver's total sensory engagement with the environment. The convergences are striking, especially bearing in mind that they came from such different intellectual backgrounds (Ingold 1996). There is no evidence that Bateson ever read or took the slightest note of the work of either Gibson or Merleau-Ponty. A comparison of their respective ideas is, however, long overdue.

sending coded messages to the brain, which it then recovers through an operation of decoding, for Bateson the world opens out to the mind through a process of revelation. This distinction, between decoding and revelation, is critical to my argument, and I shall return to it later. First, however, a few words are needed on the subject of *life*.

THE ECOLOGY OF LIFE

My leading question is one from which Bateson also set out. 'What sort of thing is this', he asked, 'which we call "organism plus environment"?' (Bateson 1973). But the answer at which I have arrived is different. I do not think we need a separate ecology of mind, distinct from the ecology of energy flows and material exchanges. We *do* however need to rethink our understanding of life. At the most fundamental level of all, we need to think again about the relation between form and process. Biology is – or at least is supposed to be – the science of living organisms. Yet as biologists gaze into the mirror of nature, what they see, reflected back in the morphology and behaviour of organisms, is their own reason. Accordingly, they are inclined to impute the principles of their science to the organisms themselves, as though each embodied a formal specification, programme or building plan, a *bio-logos*, given independently and in advance of its development in the world. The possibility of such a context-independent specification is an essential condition for Darwinian theory, according to which it is this specification, technically known as the genotype, that is said to undergo evolution through changes in the frequency of its information-bearing elements, the genes. But if the underlying architecture of the organism is pre-specified, then its life-history can be nothing more than the realization or 'writing out' of a programme of construction, under given environmental conditions. Life, in short, is regarded as purely consequential, an effect of the injection of prior form into material substance.

I take a different view of organic life, as active rather than reactive, the creative unfolding of an entire field of relations within which beings emerge and take on the particular forms they do, each in relation to the other. Life, in this view, is not the realization of pre-specified forms but the very process wherein forms are generated and held in place (Ingold 1990). A view of this kind calls for a way of thinking about evolution entirely different from the conventional Darwinian approach. For if form is a property not of genes but of the developmental dynamics of the system of relations constituted by the presence of the organism in its environment (cf. Oyama 1985), then to account for the evolution of form we have to show how such systems are constituted and reconstituted over time. In this process, organisms can play their part as producers as well as products of their own evolution, contributing through their actions to the environmental conditions both for their own development and for that of other organisms to which they relate. Every being, as it is caught up in the process and carries it forward, arises as a singular centre of awareness and agency: an enfoldment, at some particular nexus within it, of the generative potential that is life itself.

I can now spell out more precisely what I mean by an 'ecology of life'. It all hinges on a particular answer to Bateson's question: what is this 'organism plus environment'? For conventional ecology, the 'plus' signifies a simple addition of one thing to another, both of which have their own integrity, quite independently of their mutual relations. Thus the organism is specified genotypically, prior to its entry into the environment; the environment is specified as a set of physical constraints, in advance of the organisms that arrive to fill it. Indeed the ecology of the textbooks could be regarded as profoundly *anti*-ecological, in so far as it sets up organism and environment as mutually exclusive entities (or collections of entities) which are only subsequently brought together and caused to interact. A properly ecological approach is one that would take as its point of departure the organism-in-its-environment, or what phenomenological philosophers call the 'being-in-the-world'. In other words, 'organism plus environment' should denote not a compound of two things, but one indivisible totality. That totality is, in effect, a developmental system, and an 'ecology of life', in my terms, is one that would deal with the dynamics of such systems. If, in accordance with this view, we are prepared to treat form as emergent within the life-process then, I contend, we have no need to appeal to a distinct domain of mind, to creatura rather than pleroma, to account for pattern and meaning in the world. We do not, in other words, have to think of mind or consciousness as a layer of being over and above that of the life of organisms, in order to account for their creative involvement in the world. Rather, what we may call 'mind' is the cutting edge of the life process itself, the ever-moving front of what Alfred North Whitehead (1929) called a 'creative advance into novelty'.

A note on the concept of environment

Armed with this approach to the ecology of life, I shall now return to the question of how human beings perceive the world around them, in order to see how we might build an alternative to the standard anthropological account of environmental perception as a cultural construction of nature, or as the superimposition of layers of 'emic' significance upon an independently given, 'etic' reality. Before we begin, I want to make three preliminary points about the notion of environment. First, 'environment' is a relative term-relative, that is, to the being whose environment it is. Just as there can be no organism without an environment, so also there can be no environment without an organism (Gibson 1979; Lewontin 1982). Thus my environment is the world as it exists and takes on meaning in relation to me, and in that sense it came into existence and undergoes development with me and around me. Secondly, the environment is never complete. If environments are forged through the activities of living beings, then so long as life goes on, they are continually under construction. So too, of course, are organisms themselves. Thus when I spoke above of 'organism plus environment' as an indivisible totality, I should have said that this totality is not a bounded entity but a *process* in real time: a process, that is, of growth or development.

The third point about the notion of environment stems from the two I have just made. This is that it should on no account be confused with the concept of nature. For the world can exist as nature only for a being who does not belong there, and who can look upon it, in the manner of the detached scientist, from such a safe distance that it is easy to connive in the illusion that it is unaffected by his presence. Thus the distinction between environment and nature corresponds to the difference in perspective between seeing ourselves as beings *within* a world and as beings *without* it. Moreover we tend to think of nature as external not only to humanity, as I have already observed, but also to history, as though the natural world provided an enduring backdrop to the conduct of human affairs. Yet environments, since they continually come into being in the process of our lives – since we shape them as they shape us – are themselves fundamentally historical. We have, then, to be wary of such a simple expression as 'the natural environment', for in thus combining the two terms we already imagine ourselves to be somehow beyond the world, and therefore in a position to intervene in its processes.

COMMUNICATION AND REVELATION

When I was a child my father, who is a botanist, used to take me for walks in the countryside, pointing out all the plants and fungi – especially the fungi – along the way. Sometimes he would get me to smell them, or to try out their distinctive tastes. His manner of teaching was to show me things, literally to point them out. If I would but notice the things to which he directed my attention, and recognize the sights, smells, and tastes that he wanted me to experience because they were so dear to him, then I would discover for myself much of what he already knew. Now, many years later, as an anthropologist, I read about how people in Australian Aboriginal societies pass their knowledge across the generations – and I find that the principle is just the same!

In his classic study of the Walbiri of Central Australia, Meggitt describes how a boy being prepared for initiation would be taken on a 'grand tour', lasting two or three months. Accompanied by a guardian (a sister's husband) and an elder brother, the boy was taken from place to place, learning as he went about the flora, fauna, and topography of the country, while being told (by the elder brother) of the totemic significance of the various localities visited (Meggitt 1962). Every locality has its story, telling of how it was created through the earth-shaping activities of ancestral beings as they roamed the country during the formative era known as the Dreaming. Observing the waterhole while the story of its formation is related or enacted, the novice witnesses the ancestor coming out of the ground; likewise, casting his eyes over the distinctive outline of a hill or rocky outcrop, he recognizes in it the congealed form of the ancestor as it lies down to rest. Thus are truths immanent in the landscape, the truths of the Dreaming, gradually revealed to him, as he proceeds from the most superficial, 'outside' level of knowledge to deeper, 'inside' understanding.[6]

Did my father's knowledge of plants and fungi, or the Aboriginal elder's knowledge of the Dreaming, take the form of a set of interconnected beliefs and propositions inside his head? Is it through

[6]The Pintupi, whose country borders that of the Walbiri to the southwest, adopt a rather similar procedure. When young men are to be initiated, 'fathers, older brothers and other close relatives take them to sacred sites and show them rituals. They are "introduced" (*nintinu*) to the place. This visiting and seeing the site, learning about it, become important in laying claim to control or share in the control of a site' (Myers 1986). The acquisition of knowledge through initiation, whereby it passes from 'outside' to 'inside', is described as 'giving (*yunginpa*) knowledge to young people, as revealing (*yutininpa*) it, or as teaching (*nintininpa*) it' (*ibid.*).

the transfer of such beliefs and propositions from one generation to the next that we learn to perceive the world in the way we do? If so, if all knowledge is cradled within the mind, why should so much importance be placed on ensuring that novices should see or otherwise experience for themselves the objects or features of the physical world? One answer might be to suggest that it is through its inscription in such objects or features – plants and fungi, waterholes and hills – that cultural knowledge is transmitted. These objects would accordingly figure as vehicles, or carriers, for meanings that are, so to speak, 'pinned on', meanings which together constitute a specific cultural world-view or cosmology.[7] In other words, cultural forms would be encoded in the landscape just as, according to the standard semiological approach to linguistic signification, conceptual representations are encoded in the medium of sound. The great Swiss linguist Ferdinand de Saussure, who laid the foundation for this approach, argued that a sign is essentially the union of two things, a signifier and a signified, and that the relation between them is established through the mapping of one system of differences on the plane of ideas onto another system of differences on the plane of physical substance (Saussure 1959). As sounds stand for concepts, by the same logic fungi (for my father) or waterholes (for the Aboriginal elder) would stand as signifiers for elements of a comprehensive system of mental representations. Was my father, then, communicating his knowledge to me by encoding it in the fungi? Do Aboriginal elders transmit ancestral wisdom by encoding it in hills and waterholes?

Strange as it may seem, much anthropological analysis of the cultural construction of the environment proceeds from this assumption. Yet if the idea of encoding beliefs in fungi sounds bizarre, as indeed it is, the idea of the Dreaming as a cosmology encoded in the landscape is no less so. My father's purpose, of course, was to introduce me to the fungi, not to communicate by way of them, and the same is true of the purpose of Aboriginal elders in introducing novices to significant sites. This is not to deny that information may be communicated, in propositional or semi-propositional form, from generation to generation. But information, in itself, is not knowledge, nor do we become any more knowledgeable through its accumulation. Our knowledgeability consists, rather, in the capacity to situate such information, and understand its meaning, within the context of a

[7]Peter Wilson takes this view when he writes that 'the hunter/gatherer pins ideas and emotions on to the world as it exists: the landscape is turned into a mythical topographical map, a grid of ancestor tracks and sacred sites, as is typical among Australian aborigines . . .' (1988).

direct perceptual engagement with our environments. I contend that we develop this capacity by having things shown to us.

The idea of showing is an important one. To show something to somebody is to cause it to be seen or otherwise experienced – whether by touch, taste, smell, or hearing – by that other person. It is like lifting a veil off some aspect or component of the environment so that it can be apprehended directly. In this way, truths that are inherent in the world are, bit by bit, revealed or disclosed to the novice. What each generation contributes to the next, in this process, is an 'education of attention' (Gibson 1979). Placed in specific situations, novices are instructed to feel this, taste that, or watch out for the other thing. Through this fine-tuning of perceptual skills, meanings immanent in the environment – that is in the relational contexts of the perceiver's involvement in the world – are not so much constructed as discovered.

It could be said that novices, through their sensory education, are furnished with keys to meaning; but the metaphor of the key has to be used with some care. I do not have in mind the kind of key – analogous to a cipher – that might enable me to translate from physical signifiers to mental ideas and thereby to come into possession of the cultural knowledge of my forefathers through a reverse decoding of what they, in their turn, had encoded in the landscape. There is, indeed, a rather fundamental circularity in the notion that cultural knowledge is transmitted across generations by means of its encoding in material symbols. For without the key it is impossible for the novice to read off the cultural message from salient features of the physical world. Yet unless the message has already been thoroughly understood, it is impossible to extract the key. How can features of the landscape figure as elements of a communicative code if, in order to crack the code, you must already know what is to be communicated thereby?

When the novice is brought into the presence of some component of the environment and called upon to attend to it in a certain way, his task is not to decode it but to discover for himself the meaning that lies within it. To aid him in this task he is provided with a set of keys in another sense, not as ciphers but as clues (Ingold 1993*b*). Whereas the cipher is centrifugal, allowing the novice to access meanings that are attached ('pinned on') by the mind to the outer surface of the world, the clue is centripetal, guiding him towards meanings that lie at the heart of the world itself, but which are normally hidden behind the façade of superficial appearances. The contrast between the key as cipher and the key as clue corresponds to the critical distinction, to which I have already drawn attention, between decoding and revelation. A clue, in short, is a landmark that condenses otherwise disparate strands of experience into a unifying orientation which, in turn, opens up the world to perception of greater depth and clarity.[8] In this sense, clues are keys that unlock the doors of perception, and the more keys you hold, the more doors you can unlock, and the more the world opens up to you. My contention is that it is through the progressive acquisition of such keys that people learn to perceive the world around them.

FORM AND FEELING

When Susanne Langer gave the title *Philosophy in a new key* to her influential book on art and aesthetics (Langer 1957), she was of course using the metaphor of the key in yet another sense, here referring to a kind of register of understanding, akin to the key of musical notation. In the book, Langer contends that the meaning of art should be found in the art object itself, as it is presented to our awareness, rather than in what it might be supposed to represent or signify. If people in Western societies find this hard to grasp, it is because they are so used to treating art as somehow representative of something else (we expect every picture to have a title) that the ways in which we respond to objects or performances themselves are forever getting confused with our responses to whatever they are supposed to stand for. One way around this difficulty, Langer suggests, is to concentrate on the kind of art that – at least for Westerners – is apparently least representational, namely music. Music, surely, can stand for nothing but itself, so that an investigation of musical meaning should be able to show how meaning can reside in art as such. 'If the meaning of art belongs to the sensuous percept itself apart from what it ostensibly represents', writes Langer, 'then such purely artistic meaning should be most accessible through musical works' (1957). Pursuing this line

[8]Morphy appears to invoke the notion of the key in just this sense, in his account of the interpretation of so-called 'geometric' designs in the paintings of the Yolngu, an Aboriginal people of Northeast Arnhem Land. He invites us to consider a simple (and very common) motif: a circle with a line running into it. The first clue links the design with a familiar waterhole (the circle) into which there drains a creek (the line). The next links it to old man kangaroo, who made the waterhole by digging a well (the circle), using his tail as a digging stick (the line). And the next links it to the old man's penis (the line) emerging from the waterhole to penetrate the vagina (the circle) of a lady kangaroo who was bending over to drink. Through these successive revelations, the experiences of topography, subsistence activity, and sexual relations are all drawn together in an ongoing process of discovery (Morphy 1991).

This one bubbles:

That one yells:

Fig. 8.5 Janáček's sketches of the sounds of the waves, as he stood on the shore at the Dutch port of Flushing in 1926 (taken from his essay *The sea, the land*, in Janáček 1989).

of argument, Langer suggests that 'what music can actually reflect is . . . the morphology of feeling' (*ibid.*).

I believe that this idea can be generalized, so long as we recognize that feeling is a mode of active, perceptual engagement, a way of being literally 'in touch' with the world. The craftsman feels his raw material, as the potter feels clay or the turner feels wood, and out of that process of feeling there emerges the form of the vessel. Likewise, the orchestral musician feels, or rather watches, the gestures of the conductor, and out of that feeling comes a phrase shaped in sound. Thus *art gives form to human feeling*; it is the shape that is taken by our perception of the world, guided as it is by the specific orientations, dispositions, and sensibilities that we have acquired through having had things pointed out or shown to us in the course of our sensory education.

Since we are on the subject of music, let me give you one example of what I mean, taken from an essay by my favourite composer, Leoš

Janáček. Here, Janáček writes of how, on one occasion, he stood on the seashore and notated the sounds of the waves. The waves 'shout', 'bubble', and 'yell' (Janáček 1989). Figure 8.5 is a reproduction of what he put in his notebook. These musical sketches are no mere mechanical record of the sounds as they impinged on his ears. For Janáček is not just hearing, he is listening: his perception is grounded in an act of attention. Like watching and feeling, listening is something people do. In his act of attention, the movement of the composer's consciousness resonates with the sounds of the waves, and each sketch gives form to that movement.

But Janáček teaches us something more. Throughout his career, he was a compulsive collector of what he called 'speech-melodies'. He scribbled down the melodic form of snippets of speech heard from all kinds of people in all manner of activities: a housekeeper calling to her chickens as she scatters grain, an old man grumbling as he goes to work, children at play, and so on. These jottings were not confined to human sounds. Speech, for Janáček, was a kind of song, and so were all the other sounds that resonate with our consciousness, from the noises of the waves, through the tolling of an old rusty bell, or the ominous sound of a burst water-pipe, to the clucking of hens in the farmyard and the 'bloodthirsty nocturne' of a mosquito.[9] Are we to suppose, then, that in these melodies, nature is trying to communicate with us, to send us messages encoded in patterns of sound? Janáček's point was quite the opposite. It was that we should cease thinking of the sounds of speech merely as vehicles of symbolic communication, as serving to give outward expression to inner states such as beliefs, propositions, or emotions. For sound, as Janáček wrote, 'grows out of our entire being . . . *There is no sound that is broken away from the tree of life*' (1989, original emphasis).

According to Janáček, the waves shout and yell. So, sometimes, do people. When you yell in anger, the yell *is* your anger, it is not a vehicle that carries your anger. The sound is not broken off from your mental state and despatched like a message in a bottle cast upon the ocean of sound in the hope that someone might pick it up. The echoes of the yell are the reverberations of your own being as it pours forth into the environment. The philosopher Merleau-Ponty, in his *Phenomenology of perception*, caught the point precisely in his observation that your yell '*does not make me think of anger*, it is anger

[9]These examples are taken from Janáček's *Uncollected essays on music*, edited by Mirka Zemanová (1989). In letters to the great love of his life, Kamila Stösslová, Janáček jotted down the snort of his favourite domestic pig, but confessed to having some difficulty with notating the 'strange melody' of the cock that woke him every morning (Tyrrell 1994).

itself' (1962, original emphasis). If people pour out their being in the melodies of speech, so the waves pour out theirs in the sounds we describe as foaming and crashing, and the hens pour out theirs in their endless clucking. Taking one more idea from Janáček (1989), song – any song, any singing – 'is something from which we are to learn the truth of life'. This is why Aboriginal people sing their songs of the Dreaming, songs which give form to their feeling for the country around them.

CONCLUSION: TOWARDS A SENTIENT ECOLOGY

I have not forgotten the Cree hunter and the caribou, and to close my argument I want to return to them. The hunter, let us say, can *tell*. He can do so in two ways. First, he is a perceptually skilled agent, who can detect those subtle clues in the environment that reveal the movements and presence of animals: thus he can 'tell' where the animals are. Secondly, he is able to narrate stories of his hunting journeys, and of his encounters with animals. But in doing so, in telling in this other sense, he is no more aiming to produce a record or transcription of what happened than was Janáček, when he wrote down the sounds of the waves. When the hunter speaks of how the caribou presented itself to him, he does not mean to portray the animal as a self-contained, rational agent whose action in giving itself up served to give outward expression to some inner resolution. Like music, the hunter's story is a performance; and again like music, its aim is to give form to human feeling – in this case the feeling of the caribou's presence as a being-in-the-world. At that crucial moment of eye-to-eye contact, the hunter felt the overwhelming presence of the animal; he felt as if his own being were somehow bound up or intermingled with that of the animal – a feeling tantamount to love and one that, in the domain of human relations, is experienced in sexual intercourse. In telling of the hunt he gives shape to that feeling in the idioms of speech.

In a recent doctoral dissertation on reindeer herders and hunters of the Taimyr region of northern Siberia, David Anderson (1995) writes that in their relations with animals and other components of the environment, these people operate with a *sentient ecology*. This notion perfectly captures the kind of knowledge people have of their environments that I have been trying to convey. It is not knowledge of a formal, authorized kind, transmissible in contexts outside those of its practical application. On the contrary, it is knowledge based on feeling, consisting in the skills, sensitivities, and orientations that have developed through long experience of conducting one's life in a particular environment. Hunters know animals in the way we might say we 'know' other people: to know someone, in this sense, is 'to be familiar with that person's past history and sensible to his tastes, moods, and idiosyncrasies' (Ingold 1994). This is the kind of knowledge that Janáček claimed to draw from attending to the melodic inflections of speech; hunters draw it from similarly close attention to the movements, sounds, and gestures of animals.

Another word for this kind of sensitivity and responsiveness is *intuition*. In the tradition of Western thought and science, intuition has had a pretty bad press: compared with the products of the rational intellect, it has been widely regarded as knowledge of an inferior kind. Yet it is knowledge we all have; indeed we use it all the time as we go about our everyday tasks (Dreyfus and Dreyfus 1986). Moreover it constitutes a necessary foundation for any system of science or ethics. Simply to exist as sentient beings in a world, people must already be situated in a determinate environment and committed to the relationships that this entails (see Merleau-Ponty 1962). These relationships, and the sensibilities built up in the course of their unfolding, underwrite our capacities of judgement and skills of discrimination, and scientists – who are human too – depend on these capacities and skills as much as the rest of us. That is why the sovereign perspective of abstract reason, upon which Western science lays its claim to authority, is practically unattainable: an intelligence that was completely detached from the conditions of life in the world could not think the thoughts it does. It is also why reasoning logically from first principles will not suffice to design an ethical system that actually works. For any judgement that had no basis in intuition, however justified it might be on grounds of 'cold' logic, would carry no practical or motivational force whatever. Where the logic of ethical reasoning, setting out from first principles, leads to results that are counter-intuitive, we do not reject our intuitions but rather change the principles, so that they will generate results which conform more closely to what we *feel* is right.

Intuitive understanding is not contrary to science or ethics, nor does it appeal to instinct rather than reason, or to supposedly 'hardwired' imperatives of human nature. On the contrary, it rests in perceptual skills that emerge, for each and every being, through a process of development in a historically specific environment. These skills provide a necessary grounding for any system of science or ethics that would treat the environment as an object of its concern. The sentient ecology is thus both pre-objective and pre-ethical. I have no wish to devalue the projects of either natural science or environmental ethics, indeed both are probably more needed now than ever before. My plea is simply that we should not lose sight of their pre-objective, pre-ethical foundations. My overriding aim has been to bring these foundations to light. What these excavations into the formation of knowledge have revealed is not an alternative science, 'indigenous' rather than Western, but something more akin to a *poetics of dwelling*. I contend that it is within the framework of such a poetics that Cree tales of animals offering themselves to humans, Aboriginal stories of ancestors emerging from waterholes, Janáček's attempts to notate the sounds of nature, and my own father's efforts to introduce me to the plants and fungi of the countryside, can best be understood.

REFERENCES

Anderson, D. (1995). *National identity and belonging in Arctic Siberia*. Unpublished Ph.D. thesis, University of Cambridge.

Bateson, G. (1973). *Steps to an ecology of mind*. Granada, London.

Bateson, G. (1980). *Mind and nature: a necessary unity*. Fontana/Collins, London.

Brightman, R. (1993). *Grateful prey: Rock Cree human–animal relationships*. University of California Press, Berkeley.

Dreyfus, H. and Dreyfus, S. (1986). *Mind over machine*. Free Press, New York.

Feit, H. (1973). The ethno-ecology of the Waswanipi Cree: or how hunters can manage their resources. In *Cultural ecology: readings on the Canadian Indians and Eskimos* (ed. B. Cox), pp. 115–25. McClelland and Stewart, Toronto.

Gibson, J.J. (1979). *The ecological approach to visual perception*. Houghton Mifflin, Boston.

Gudeman, S. (1986). *Economics as culture*. Routledge and Kegan Paul, London.

Ingold, T. (1980). *Hunters, pastoralists and ranchers: reindeer economies and their transformations*. Cambridge University Press.

Ingold, T. (1990). An anthropologist looks at biology. *Man* (N.S.), **25**, 208–29.

Ingold, T. (1993a). The art of translation in a continuous world. In *Beyond boundaries: understanding, translation and anthropological discourse* (ed. G. Pálsson), pp. 210–30. Berg, Oxford.

Ingold, T. (1993b). The temporality of the landscape. *World Archaeology*, **25**(2), 152–74.

Ingold, T. (1994). From trust to domination: an alternative history of human-animal relations. In *Animals and human society: changing perspectives* (ed. A. Manning and J. Serpell), pp. 1–22. Routledge, London.

Ingold, T. (1996). Culture, perception and cognition. In *Psychological research: innovative methods and strategies* (ed. J. Haworth), pp. 99–119. Routledge, London.

Janáček, L. (1989). *Janáček's uncollected essays on music* (trans. and ed. M. Zemanová). Marion Boyars, London.

Langer, S. (1957). *Philosophy in a new key*. Harvard University Press, Cambridge, Mass.

Lévi-Strauss, C. (1974). Structuralism and ecology. *Social Science Information*, **12**(1), 7–23.

Lewontin, R. C. (1982). Organism and environment. In *Learning, development and culture* (ed. H.C. Plotkin), pp. 151–70. John Wiley, Chichester.

Mech, L.D. (1970). *The wolf.* Natural History Press, Garden City.

Meggitt, M.J. (1962). *Desert people: a study of the Walbiri Aborigines of Central Australia*. Angus and Robertson, Sydney.

Merleau-Ponty, M. (1962). *Phenomenology of perception* (trans. C. Smith). Routledge and Kegan Paul, London.

Merleau-Ponty, M. (1964). Eye and mind. In *The primacy of perception and other essays on phenomenological psychology and the philosophy of art, history and politics* (ed. J.M. Edie), pp. 159–90. Northwestern University Press, Evanston, Ill.

Morphy, H. (1991). *Ancestral connections: art and an Aboriginal system of knowledge*. University of Chicago Press, Chicago.

Myers, F.R. (1986). *Pintupi country, Pintupi self: sentiment, place, and politics among Western Desert Aborigines*. Smithsonian Institution Press, Washington.

Oyama, S. (1985). *The ontogeny of information: developmental systems and their evolution*. Cambridge University Press.

Saladin d'Anglure, B. (1979). *Inuit and caribou* (Inuit texts and illustrations concerning caribou, collected by B. Saladin d'Anglure). Inuksiutiit Allaniagait 2, Association Inuksiutiit Katimajiit, Départment d'Anthropologie, Université Laval, Quebec, Canada.

Saussure, F. de (1959). *Course in general linguistics*. Philosophical Library, New York.

Scott, C. (1989). Knowledge construction among Cree hunters: metaphors and literal understanding. *Journal de la Société des Americanistes*, **75**, 193–208.

Tanner, A. (1979). *Bringing home animals: religious ideology and mode of production of the Mistassini Cree hunters*. St Martin's Press, New York.

Tyrrell, J. (1994). *Intimate letters: Leoš Janáček to Kamila Stösslová* (ed. and trans. J. Tyrrell). Faber and Faber, London.

Whitehead, A.N. (1929). *Process and reality: an essay in cosmology*. Cambridge University Press.

Wilson, P.J. (1988). *The domestication of the human species*. Yale University Press, New Haven.

Glossary

agonist Muscle whose active contraction causes movement of a part of the body.

alienation In psychiatry the thought that one's thoughts are under the control of somebody else, or that other people participate in one's thinking.

amygdala A roughly almond-shaped mass of grey matter deep inside the temporal lobe of each cerebral hemisphere. It has extensive connections with the olfactory system and sends fibres to the hypothalamus: its functions are apparently concerned with mood, feeling, instinct, and possibly memory for recent events.

amyloid (plaques) Glycoprotein, resembling starch that is deposited in the internal organs in amyloidosis.

astrogliosis Astrocyte – one of the several different types of cell making up the glia. The cells have been ascribed the function of providing nutrients for neurones and possibly of taking part in information storage processes. To do with the brain becoming spongy.

ataxia Shaky movements and unsteady gait that result from the brain's failure to regulate the body's posture and the strength and direction of limb movements.

atrophy The wasting away of a normally developed organ or tissue due to degeneration of cells.

autosomal dominant Autosome – any chromosome that is not a sex chromosome and occurs in pairs in duploid cells means that the disease is genetically determined.

axon A nerve fibre: a single process

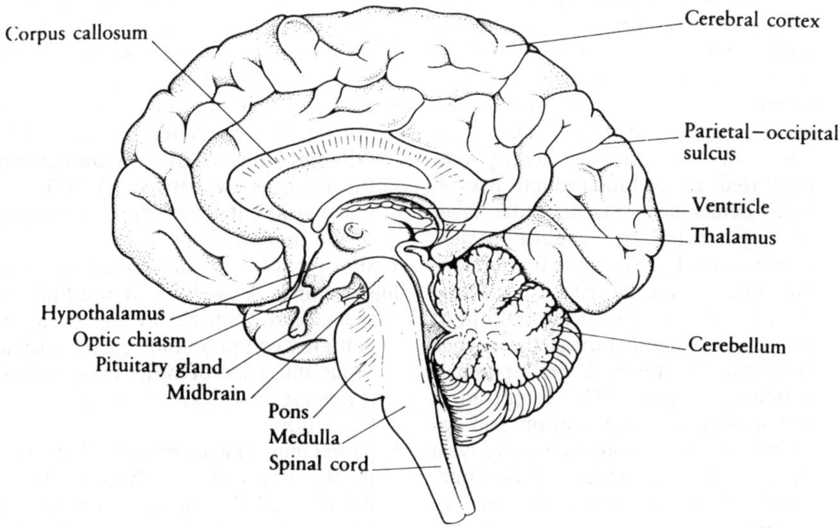

Fig. A1 Mid-sagittal section of human brain (a vertical section taken between the eyes).

extending from the cell body of a neuron and carrying nerve impulses away from it.

brainstem An enlarged extension upwards of the spinal cord, within the skull, consisting of the medulla oblongata, the pons, and the midbrain. The pons and medulla are together known as the bulb or bulbar area

cerebellum The largest part of the hindbrain. Like the cerebrum, it has an outer grey cortex and a core of white matter. Three broad bands of nerve fibres connect it to the medulla, the pons, and the midbrain. It has two hemispheres. It is essential for the maintenance of muscle tone, balance, and the synchronization of activity in groups of muscles under voluntary control, converting muscular contractions into smooth movements.

cerebral cortex The intricately folded outer layer of the cerebrum, making up some 40% of the brain by weight and composed of an estimated 15 thousand million neurones. This is the part of the brain most directly responsible for consciousness, with essential roles in perception, memory, thought, mental ability, and intellect, and it is responsible for initiating voluntary activity.

cerebrum The largest and most highly developed part of the brain, composed of two cerebral hemispheres separated by a longitudinal fissure. Each hemisphere has an outer layer of grey matter, the cerebral cortex, below which lies the white matter containing the basal ganglia. Connecting the two hemispheres at the bottom of the longitudinal fissure is the corpus callosum. Within each hemisphere is a lateral ventricle. The cerebrum is responsible for the initiation and co-ordination of all voluntary activity in the body and for governing the functioning of lower parts of the nervous system.

corpus callosum The broad band of nervous tissue that connects the two cerebral hemispheres, containing an estimated 300 million fibres.

dendrite Short branch of a neuron, which receives contacts from the axons of other neurons at synapses and carries signals from them into the cell body.

dura mater The thickest and outermost part of the three meninges surrounding the brain and spinal cord.

epithelium The tissue that covers the external surface of the body and lines hollow structures. In the brain, the progenitor epithelium gives rise to all the neurons.

exteroceptor A sensory nerve, ending in the skin or mucous membrane, that is responsive to stimuli from outside the body.

ganglion; basal ganglia Any structure containing a collection of nerve cell bodies and often also numbers of synapses. Within the CNS certain well-defined masses of nerve cells are called ganglia e.g. basal ganglia – several large masses of grey matter embedded deep within the white matter of the cerebrum. They have complex neural connections with both the cerebral cortex and the thalamus: they are involved with the regulation of voluntary movements at a subconscious level.

glia The special connective tissue of the central nervous system, composed of different cells. Glial cells outnumber the neurons by between 5–10:1 and make up about 40% of the total volume of the brain and spinal cord.

GSR A type of skin conductance response.

hippocampus A swelling in the floor of the lateral ventricle of the brain. It contains complex foldings of cortical tissue and is part of the limbic system. It is involved in certain kinds of memory formation.

hypothalamus The region of the forebrain in the floor of the third ventricle, linked with the thalamus above and the pituitary gland below. It contains important centres for controlling body temperature, thirst, hunger, eating, water balance, and sexual function. It is closely connected with emotional activity and sleep.

iatrogenic A condition that has resulted from treatment, as either an unforeseen or inevitable side-effect.

interoceptor Any receptor organ composed of sensory nerve cells that respond to and monitor changes within the body, such as the stretching of muscles.

limbic system The amygdala, hippocampal formation, and hypothalamus. The activities governed are those concerned with self-preservation, preservation of the species, and the expression of rage, fear, pleasure and the establishment of memory patterns.

lobe Major division of an organ, especially one having a rounded form and often separated from other lobes by fissures or bands of connective tissue

mitochondria Occur in all cells, at the site of the cell's energy production. They contain enzymes involved with the cell's metabolic activities

neuron Nerve cell – one of the basic functional units of the nervous system. The cells are specialized to transmit electrical nerve impulses and carry information from one part of the body to another. Each contains a nucleus, *dendrites* (through which signals from other neurons enter the cell), and an *axon* – a longer process that extends out and carries impulses away from the body. The point of contact of one neuron with another is called a *synapse*.

NMDA receptors *N*-Methyl-D-aspartate receptors are activated under specific conditions and tend to strengthen the efficiency of that synapse

penetrance The frequency with which the characteristic controlled by a gene is seen in the individuals possessing it. Complete penetrance occurs when the characteristic is seen in all individuals known to possess the gene. If many individuals with the gene do not show its effects, penetrance is said to be incomplete. In this way a trait may appear to skip a generation.

pons Varolii The part of the brainstem that joins the medulla oblongata and the thalamus, bulging forwards in front of the cerebellum from which it is separated by the fourth ventricle.

prefrontal lobe The region of the brain at the very front of each cerebral hemisphere. The functions of the lobe are concerned with emotions, memory, learning, and social behaviour.

prion A small proteinaceous infectious particle which resists inactivation by procedures that modify nuclear acids.

prosopagnosia The condition of a brain-injured person who is still able to see but has lost the ability to recognize people from their faces; the ability to recognize people from other clues, such as voice or name, is unaffected.

pyramidal cell A type of neuron found in the cerebral cortex and elsewhere, with a pyramidal-shaped cell body, a branched dendrite extending up from the apex towards the brain surface, several dendrites extending horizontally from the base, and an axon running in the white matter of the hemisphere.

receptor (1) A cell or group of cells specialized to detect changes in the environment and trigger impulses in the sensory nervous system. All sensory nerve endings act as receptors, whether they simply detect touch, as in the skin, or chemical substances as in the nose and tongue or sound or light, as in the ear and eye.

(2) A complex protein molecules embedded in membrane of a nerve cell.

SCR Skin conductance response; same as GSR.

somatic Soma – the body, as distinct from the mind.

Somatic sensory area – strip of cortex in the middle of the hemisphere.

Somatosensory structures – regions of brain which receive and map body signals.

spatial contrast sensitivity function Visual stimuli can be broken down mathematically into signals varying from high to low spatial frequencies. The spatial contrast sensitivity function charts this by measuring the lowest contrast at which patterns of different spatial frequency can be detected.

synapse The minute gap across which nerve impulses pass from one neuron to the next, at the end of a nerve fibre. Reaching a synapse, an impulse causes the release of a neurotransmitter, which diffuses across the gap and triggers an electrical response in the next.

temporal lobe One of the main divisions of the cerebral cortex in each hemisphere of the brain, lying at the side within the temple of the skull and separated from the frontal lobe by a cleft. Areas of the cortex in this lobe are concerned with memory and the appreciation of sound and spoken language.

thalamus One of two egg-shaped masses of grey matter that lie deep in the cerebral hemispheres in each side of the forebrain. The thalami are relay stations for sensory messages that enter the brain, transmitting them to the cortex. All sensory pathways except for that of smell, are linked to nuclei within the thalamus.

ventricle One of four fluid-filled cavities in the brain. The paired first and second ventricles (lateral), one in each cerebral hemisphere, communicate with the third ventricle in the midline between them. This in turn leads through a narrow channel to the fourth ventricle in the hindbrain which is continuous with the spinal canal in the centre of the spinal cord.

viroids Variation on a virus.

Index